New England Weather, New England Climate

Gregory A. Zielinski

Barry D. Keim

New England Weather, New England Climate

University of New Hampshire

Published by University Press of New England

Hanover and London

University of New Hampshire

Published by University Press of New England, 37 Lafayette St., Lebanon, NH 03766

© 2003 by Gregory A. Zielinski and Barry D. Keim

Printed in the United States of America

5 4 3 2 1

Library of Congress Cataloging-in-Publication data

Zielinski, Gregory A.
    New England weather, New England climate / Gregory A. Zielinski,
Barry D. Keim.
        p.   cm.
    Includes bibliographical references and index.
    ISBN 1-58465-312-4 (cloth : alk. paper)
        1. New England—Climate. I. Keim, Barry D., 1963– II. Title.
QC984.N35 Z54 2003
551.6974—dc21   2002152726

Contents

Figures and Tables

Figures

Tables

Acknowledgments

We would like especially to thank the University Press of New England for suggesting to GAZ that a revision and expansion of David Ludlum's book on New England weather was very much needed. We took his suggestion and produced this compilation of how the weather and climate of the region works. We were able to take time from our other research and teaching activities to put this work together thanks, in part, to two funding sources to the Climate Change Research Center when we both were part of that group. The initial work on this book was partially supported by the Iola Hubbard Climate Change Endowment for the Climate Change Research Center as established by Leslie S. Hubbard in memory of his late wife. Partial support was also provided through funding from the National Oceanic and Atmospheric Administration (NOAA) through the AIRMAP program at the Climate Change Research Center. The final stages of this effort were partially supported by a National Science Foundation grant to GAZ from the Paleoclimate Program to evaluate the historical climatology of New England. Coversations with many other individuals have helped put ideas into our heads regarding those aspects of New England weather and climate that are important to the people of the region. Some of these are Dave Thurlow, Ernie and Shawn Roberts, Susan Jasse, Bob Moulton, Roy Hutchinson, Alden Marshall, Dick Dionne, Northan Parr, Carl and Carol Wilden, James Cerny, Ava Strait, and Norman Macdonald. We apologize to those whom we may have forgotten.

In addition to the individuals and organizations that provided assistance in putting this book together, we are both greatly indebted to our families and friends for tolerating the times when we were occupied by this endeavor and for encouraging us to finish it while other demands were in place. Ann Zielinski helped tremendously with many of the figures.

Greg Zielinski wishes to thank his wife, Ann, and children Chris, Catie, and Andy for their continued support and love while this book was being written. Their support was there in many ways, but they helped especially by taking up the slack in family duties that should have been done by me. Without them, this book may not have been completed. Many of my colleagues have supported this work in many ways, but I especially want to thank Paul Mayewski for giving me the opportunity to continue to pursue this work (and other research opportunities) through the position he established for me at the University of Maine. In closing, I certainly want to thank my father and late mother for their support and assistance as I pursued my career.

Barry Keim thanks his wife Ellen, and their two sons, Anderson and Nathan, for their love, support, and understanding throughout the writing of this book. They have sacrificed in ways far too numerous to mention. I also lovingly acknowledge my parents, Mr. and Mrs. Elwood H. Keim, for providing me a stable home and for making sacrifices so that I could pursue a career in academics.

PART I

What Makes New England's Weather and Climate Unique

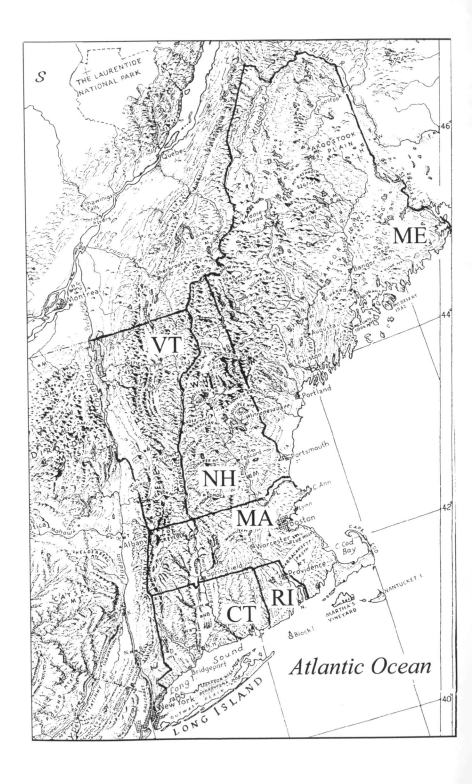

Chapter I

A General Perspective

Yes, one of the brightest gems in the New England weather is the dazzling uncertainty of it. —MARK TWAIN

Weather and Climate as a Way of Life

It is early morning, mid-week. The rain is coming down heavily as you sit at the breakfast table contemplating not how the rain will affect you today, but whether it will clear up for that special event over the weekend. These thoughts are interrupted occasionally by thoughts of whether you should plan to start that big project around the house this spring, remembering the very warm and dry March and April of last year. Unfortunately, in the back of your mind, you know that this March, and even this April, may have a heavy snowfall almost anytime during the month. Ah, what to do?

This common scenario happens quite frequently to just about everyone. If you live in New England or have visited the region, you know that your question is probably not an easy one to answer, since New England's weather is arguably one of the most varied in the world over such a small area (fig. 1.1). Not only is the weather quite variable from season to season, but it includes extremes of both hot and cold temperatures, droughts, heavy rainfall, hurricanes, tornadoes, blizzards, and more. This high level of changeability over very short periods of time accompanies abrupt and distinct changes in weather and climatic conditions over short distances across the region. New Englanders say, "If you don't like the weather, wait a minute." In New England, one could

Fig. 1.1. (*opposite*) Physiographic map of the six New England states (66.9° to 73.6°W, 41.0 to 47.5°N), as well as parts of the surrounding states and Canadian provinces, showing the features on the landscape that have a major influence on New England's climate. These include the interior mountains and the Atlantic Ocean. Names of larger cities and towns may be abbreviated by the first letter of the name (such as, #H for Hartford, Connecticut, #M for Manchester, New Hampshire, and #B for Burlington, Vermont). State abbreviations are used to reduce masking of physiographic features. Modified from map entitled *Landforms of the United States* by Erwin Raisz (6th revised edition, 1957).

also argue that if you don't like the weather, drive a few hundred yards. Not only does the weather here change quickly, but it also changes spatially.

Although many parts of the United States make this same claim, we will demonstrate in this book that perhaps nowhere is this statement more true than in New England. In fact, to get a quick perspective on how variable New England's climate is over a short distance, all one has to do is to look at the changes in plant hardiness zones across New England compared to the those in the rest of the country (fig. 1.2). There are five hardiness zones in New England, Zones 3 to 7, more than any equivalent size region in the country. Only easternmost New York goes through five zones over an equal distance. As gardeners throughout New England know, one does not have to travel far to enter a different plant hardiness zone or subzone (fig. 1.3).

Before we delve too deeply into the subject, a quick clarification of the terms "weather," "meteorology," and "climate" is in order. "Weather" represents what is occurring in the atmosphere (e.g., temperature, precipitation, wind, relative humidity) at any one point in time or over very short periods of time at some specific location (usually locally), mainly with respect to effects upon life and human activities (Schneider, 1996; Glickman, 2000). In essence, the weather is what you talk about in line at the supermarket. As a result, dif-

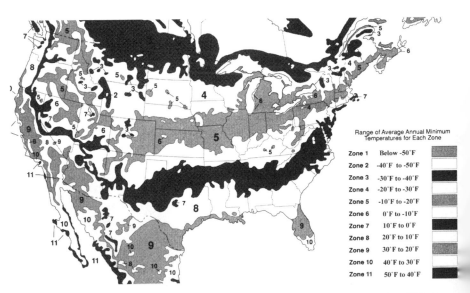

Fig. 1.2. Plant hardiness zones across the contiguous United States and parts of Canada and Mexico. Note the number of zones across New England compared to other parts of the country. The projection of this map makes the states of New England look larger than they actually are compared to other states. Modified from *The Big Book of Gardening Skills* by the editors of Garden Way Publishing (p. 168), copyright © 1993 by Storey Communications, Inc. Reprinted with permission from Storey Publishing, LLC.

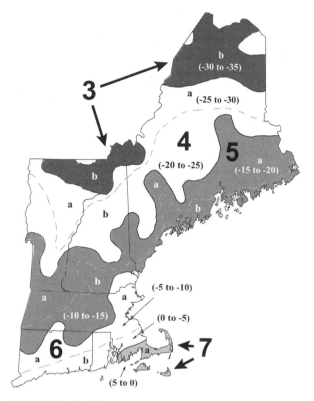

Fig. 1.3. Plant hardiness zones and subzones in New England as modified from figure 1.2 and the 2001 U.S. National Arboretum "Web Version" of the USDA Plant Hardiness Zone Map (USDA, 1990). A colored version is available at http://www.usna.usda.gov/Hardzone/ushzmap.html, at the time this was written. Dashed lines are approximate boundaries for individual subzones as marked by an a or b. Subzones further divide the range of average annual minimum temperatures into 5°F ranges as indicated.

ferent types of weather can occur simultaneously, even at places that are very near to each other. For example, Mount Washington's weather may consist of high winds and snow, while at the same time North Conway, New Hampshire, only about 20 miles away, is sunny and mild.

"Meteorology" is the science of weather and the study of atmospheric phenomena. In popular usage, it is the information concerning fronts and the movements of weather patterns that are reported on television or radio to explain and predict weather conditions tomorrow and the next few days to come (Glickman, 2000).

In contrast, "climate" represents average conditions over longer time periods or over larger areas (or both). For example, the mean January temperature or mean annual precipitation in Boston or in Massachusetts represents climate information. Climate also includes the rates of occurrence of extreme events, answering questions such as "what are the chances of New England experiencing a hurricane in a given year?" Climatology is the description and study of climate and it can also include forecasting, but forecasting over the long term, such as for the upcoming thirty days or season or year, or even longer.

New Englanders often have a love-hate relationship with the high level of variability of their weather and climate. New Englanders tend to wax poetic about their multi-season climate, although they become agitated when certain kinds of weather take place in the wrong season. For example, winter's snowstorms should take place in December, January, and February. Everyone accepts this, but when snowstorms occur in mid-April or mid-October, everyone gets up in arms. Furthermore, summer's high temperatures and humidity were meant for summer, but these conditions sometime sneak up on you in March, or they may wait until December. However, given that New England winters have teeth, even the heartiest of residents become anxious when spring is slow to manifest itself. Mark Twain may have addressed this situation best:

The people of New England are by nature patient and forbearing, but there are some things which they will not stand. Every year they kill a lot of poets for writing about Beautiful Spring. These are generally casual visitors, who bring their notions of spring from somewhere else, and cannot, of course, know how the natives feel about spring. (Twain, 1935)

The fickle nature of New England weather makes it a prime topic of conversation, and perhaps even a way of life in the region. After all, many segments of the northeastern economy are dependent on the weather, and most people would like it to behave in a civilized manner . . . or so they claim. Again, Mark Twain expressed this feeling in one of his speeches about New England's weather. "Yes, one of the brightest gems in the New England weather is the dazzling uncertainty of it. There is only one thing certain about it: you are certain there is going to be plenty of it—a perfect grand review; but you never can tell which end of the procession is going to move first." Twain further elaborated on the splendors of New England weather and its bountiful variety in his speech at the New England Society's seventy-first annual dinner in New York. The following excerpt may say it best.

Now as to the size of the weather in New England—lengthways, I mean. It is utterly disproportionate to the size of that little country. Half the time, when it is packed as full as it can stick, you will see New England weather sticking out beyond the edges and projecting around hundreds and hundreds of miles over the neighboring states. She can't hold a tenth part of her weather. (Twain, 1935, p. 1110)

Twain nailed New England's weather and climate on the head with his tremendous wit, and we acknowledge him by using many of his quotations as epigraphs to the various sections and chapters in this book.

Despite the richness of the weather here, the attention given to New England weather by poets (such as Robert Frost) and satirists, and the abundance of severe weather types, few books have focused on understanding the dynamic climate of the region. One of the best efforts to date in this regard was written by David Ludlum (1976) in his *The Country Journal: New England Weather Book*. He compiled many fascinating statistics to demonstrate the extreme range of weather across the region. However, many impressive weather events have occurred since the publication of Ludlum's book. Furthermore, issues of climate change and its impacts on the region have risen to the forefront of scholarship as well as to the minds of government agencies and officials interested in the future socioeconomic welfare of their consitutents. It is our intent to broaden the understanding of this region's interesting and diverse weather and climate, while addressing issues of climate change and its importance to New Englanders. Figure 1.1 shows the diverse New England landscape and the location of several cities within the region relative to the interior mountains and the coastline, the importance of which we elaborate on throughout this book.

What Follows

We divide the book into seven general parts. Part I, "General Perspective," consists of this introduction to how New England's weather and climate is perceived by individuals who live in or visit the region. Part II, entitled "Causes of Change in New England's Weather and Climate," contains three chapters that provide information on different time scales. Chapter 2 explains what causes seasonal changes within the year. This is followed by a discussion of the factors that cause the region's weather and climate to change from year to year (chapter 3). Chapter 4 presents the factors that cause changes on longer time scales of decades, centuries, millennia, and beyond.

Part III, "Diversity of New England's Weather and Climate: From the Mountains to the Beaches," covers the factors that cause changes spatially across the region. Chapter 5 presents the permanent causes for different climatic zones across New England, and Chapter 6 presents those causes that can vary on different time scales.

Part IV, "Seasons of New England," contains six chapters that discuss the characteristics of the "seasons" of New England. We do not describe the typical solar seasons of spring (March, April, May), summer (June, July, August), autumn (September, October, November), and winter (December, January,

February). Rather, we define the seasons by the manner in which they influence society. Consequently, the timing covered in each chapter differs slightly from the typical seasons stated above. Chapter 7 presents the many characteristics of ski season. Chapter 8 expands on the various aspects of mud season. Beach and lake season are addressed in chapter 9. Chapter 10 discusses foliage season. A summary of the year is given in chapter 11, and the special weather and climatic conditions of New England's mountains, the Alpine Zone, are described in chapter 12.

Part VII also contains six chapters, but the emphasis now is on meteorological or climatological events that impact the everyday lives of New Englanders: "Events that Influence the Lives of New Englanders." Chapter 13 includes information on temperature-related events, while events related to the precipitation component of New England's climate system are covered in chapter 14. The next four chapters discuss four different types of events that truly show the diversity of New England's weather and climate: chapter 15 covers nor'easters, chapter 16 is on ice storms, chapter 17 provides information on tornadoes, and chapter 18 discusses hurricanes.

Part VI of our book examines an aspect of New England's climate that has not been evaluated in great detail, despite the information available. The two chapters in Part VI cover "Changes over Time." Chapter 19 describes how our climate has changed in the recent past up until the present (that is, over the last three to four centuries since European settlement). Chapter 20 then looks into the future to create a possible scenario of what may happen to our weather and climate as we progress through this new millennium.

The last part of the book, "A Retrospective," provides a general summary of the preceding parts of our work. This is followed by a list of references cited, a glossary of some of the more important and uncommon terms that we use, and a list of additional readings that may be of interest to aficionados of weather and climate, as well as individuals who found that our book sparked an interest in weather and climate. If our book does this for just one individual, we will consider it a success.

PART II

Causes of Change in New England Weather and Climate

If you don't like the weather, wait a minute.   —ANONYMOUS

Although many areas of the country make the claim that one only has to wait a minute and the weather will change, perhaps no other region of the same size truly has such variability. Even more interesting is the fact that the weather can change drastically not only from day to day, but over monthly and yearly time scales. There may be no better example of this scenario than the meteorological conditions in Boston on 31 March and 1 April in 1997 and 1998. On 31 March 1997, it began to snow in Boston late in the afternoon, and did not stop until late in the day on 1 April. The April Fools' snowstorm of 1997 produced 25.4 inches near Logan Airport, becoming the twenty-four-hour snowfall record for Boston. Temperatures during the snowstorm were in the low 30s. The very next year, a high temperature of 89°F was recorded on 31 March at Logan Airport. Reading, Massachusetts, recorded 92°F during this same event, which stands as the all-time highest temperature ever recorded in the month of March in the state of Massachusetts. This high temperature of 92°F on 31 March 1998 was 45°F warmer than the maximum temperature of 47°F on 31 March 1997. Although these year-to-year changes may be quite extreme, they pale in comparison to changes that can occur over longer time periods, such as from century to century and especially from millennium to millennium. Imagine what New England was like just 20,000 years ago—a short period of time from a geological perspective. The region was under three miles of ice as the large ice sheets that formed on the continents of the northern hemisphere dominated global weather and climate. In fact, just 10,000 years ago, these ice sheets still may have covered parts of New England. The landscape adjacent to the glacier was covered by an abundance of sediment left by the ice and was void of much of the vegetation we see today.

Considering how rapidly changing and variable weather and climate conditions are in New England and how quickly these changing conditions can impact humans, it is important to understand why such rapid changes occur. In this part of the book we will answer the question "What are the factors that force New England's climate to be so variable over these different periods of time?" The general answer is that there are many different factors, and which particular factors come into play is a function of the time period discussed. For instance, the factors that cause variability in weather and climate from day to day and season to season are much different than those that cause changes over hundreds of years, as well as over thousands of years. The next three chapters introduce the permanent factors that cause changes in New England's weather and climate. Chapter 2 describes what causes the day-to-day changes *that ultimately lead to changes from season to season. We first introduce specific factors that result in changes throughout the day, called diurnal changes. The items presented in this chapter are brief summaries of many complex meteorological phenomena, and we direct the reader to the meteorological textbooks referenced in the back for more detailed explanations of these concepts. Chapter 3 introduces factors that influence our weather and climate from year to year, as well as from decade to decade. Chapter 4 then summarizes the factors that control climate over millennial and greater time frames, the time scales in which glacial and interglacial periods are operative.*

Chapter 2

Changes within the Year

The weather is always doing something there. —MARK TWAIN

Change is the hallmark of New England weather. What better way to come to grip with it than to look at what controls these natural systems from day to day, week to week, and month to month through the year. We start by summarizing the important details that influence weather conditions during the day, followed by important factors that control these conditions throughout the year. We finish this chapter by summarizing the major circulation patterns in the atmosphere that so highly influence changes during the year.

Diurnal Controls

The most dominant factor responsible for changing atmospheric conditions not only through the seasons, but during each and every day is the variability in the amount and intensity of insolation (short for INcoming SOLar radiATION) reaching Earth's surface. There is a simple explanation why the coldest part of the day is just after the sun rises, that is, almost exactly the time on those cold winter mornings when you are leaving to go to work or school (fig. 2.1). Early morning is the coldest time of the day because of a process called radiational cooling. The land holds the daytime heat generated by insolation, but this heat is lost rapidly once the sun sets, with a continual loss of heat throughout the night. By morning, the maximum amount of heat has radiated away, and the sun has not yet begun to warm the land again. Clearer nights lead to greater radiational cooling because clouds tend to hold in the heat at night. However, when windy conditions exist at night, the resulting mixing of the air column will produce warmer temperatures at the surface than if conditions were calm. Atmospheric conditions occasionally will produce colder temperatures during a different part of the day, such as when a cold air mass moves into the region, thereby decreasing temperatures throughout the day.

A similar concept holds true for the hottest part of the day. The surface receives the most direct insolation at solar noon, when the sun is directly overhead. However, the hottest part of the day occurs about two to three hours later (fig. 2.1). This delay in heating occurs because the land continues to absorb more radiation than it loses despite the decreasing angle of the sun and the less intense solar radiation at the earth's surface as the day progresses. As is the case for the timing of the coldest part of the day, a change in air mass in the area may result in the warmest part of the day occurring at a time other than late afternoon.

The peak in daytime heating during the afternoon produces several daily phenomena in New England, primarily during the summer season. The air immediately above the surface may become unstable with continued heating throughout the day. Parcels of air will then rise and cool. As it cools, an air parcel will eventually reach the temperature of condensation (or dew point), whereby water vapor within the parcel condenses and eventually forms clouds. If the air over a region is unstable with much vertical air movement, fluffy cumulus clouds may continue to grow upward, sometimes producing cumulonimbus clouds, that is, thunderheads (see chapters 9 and 14). This is the reason that most thunderstorm activity occurs in the late afternoon or early evening during the summer.

Another daily phenomena that is primarily a summertime feature along the seacoast of New England is the formation of a sea breeze. This cooling breeze

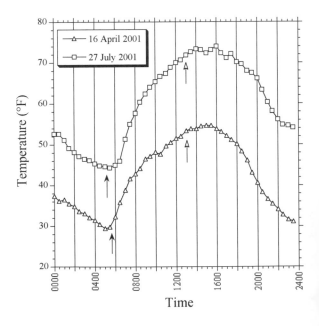

Fig. 2.1. Diurnal temperature variability for 16 April and 27 July 2001 in Bangor, Maine, showing the relationship between the coldest time of day and sunrise (solid arrows), as well as the lag time between the hottest part of the day and the time of greatest solar radiation, or solar noon (open arrows). Solar noon was at 1300 hours daylight savings time for both days.

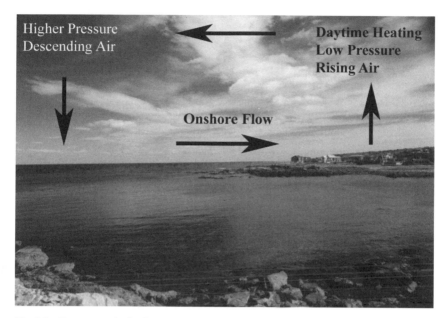

Fig. 2.2. Components in the formation of a sea breeze. Daytime heating of the land starts the process of air movement in both the vertical and horizontal directions (shown by arrows and discussed in text). The process reverses at night when the greater cooling of the land causes higher pressure and descending air over the land. Horizontal air movement is then offshore with rising air over the ocean, thereby reversing the cycle from that shown here. Photo by Greg Zielinski taken from Marginal Way, Ogunquit, Maine.

forms as the heating of the land causes air parcels to rise, creating lower atmospheric pressure over the land than is over the ocean (fig. 2.2). Air flow moves onshore, drawing cool air from over the cooler ocean waters, thereby producing cool air temperatures along the shoreline and a haven for those who want to escape the summer heat. At night, the process reverses. As the land cools, the air above it will sink, forming higher pressure over land than over ocean waters. Wind flow will then be directed offshore, because of the pressure gradient produced.

Yearly Controls

In addition to the diurnal variability in the intensity of insolation reaching Earth's surface, yearly insolation variability is responsible for New England's distinct seasons and the differing climatic conditions in each season. The cause for the seasons is the 23° 27′ tilt (at the present) of Earth's axis. The tilt produces a change in Earth's orientation to the sun as Earth revolves around the sun during the year. The sun's rays strike the northern hemisphere at higher angles, as well for more daylight hours during summer (fig. 2.3). Consequently, solar radiation reaching the northern hemisphere is more intense during the summer than during the winter. For instance, the sun is 70° above the horizon at high noon on the summer solstice at Concord, New Hampshire (90° would be directly overhead), and just 23° above the horizon on the winter solstice. When the angle of the sun is low, the sun's rays travel through more of the atmosphere, which may allow greater reaction with or deflection by dust and aerosols, decreasing the amount of insolation during the winter months. The change in position of the sun in the winter sky (closer to the horizon and farther south) compared to the sun's position in the summer sky (higher in the horizon and farther north) is visible evidence of changes in the angle of the sun's rays hitting Earth with the change in seasons (fig. 2.3). The combination of lower sun angle and shorter daylight hours means that less solar radiation reaches the northern hemisphere in winter, so there is less heating during the day and over the season as a whole. The rapid change in the number of daylight hours during the spring and fall transition seasons, together with the rapidly changing amount of insolation, produces quickly changing climatic conditions during those months (fig. 2.4; table 2.1).

The amount of solar radiation reaching the top of Earth's atmosphere across New England is almost six times greater on 20 June than on 20 December. At the spring and fall equinoxes, the region receives a little more than half of what it receives on the summer solstice. That is a huge difference in the amount of heat received at the top of the atmosphere between winter and summer. How-

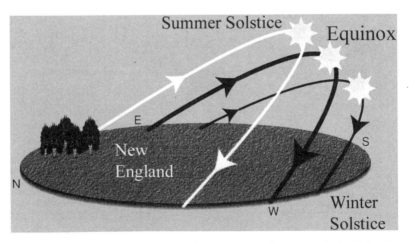

Fig. 2.3. Change in angle of the sun throughout the year as it traverses the sky during the day. Modified from Danielson et al. (1998; fig. 3.7, p. 75). Used with permission from The McGraw-Hill Companies.

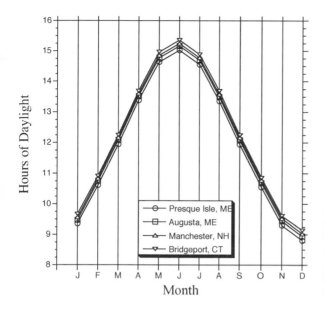

Fig. 2.4. Example of northeast to southwest variability in the number of hours of daylight across New England for the twenty-first day of each month.

ever, day-to-day variability in cloud cover, dust, and other types of particles in the atmosphere modifies the amount of that insolation reaching the surface. Aerosols will either reflect or scatter incoming solar radiation or possibly absorb the radiation. Similarly, clouds can reflect solar radiation before it hits Earth's surface. Different land conditions reflect solar radiation back into the atmosphere at various levels. For example, fresh snow can reflect up to 95 percent of insolation, whereas forest cover reflects only 5 to 20 percent, with deciduous forest at the higher end of that range and coniferous forest at the lower end. As you see, many factors affect seasonal variability in solar radiation reaching the top of Earth's atmosphere, resulting in modified climatic conditions.

Global Circulation Patterns

The distribution of insolation does not just vary with time during the day and year. It also varies regionally across Earth. A greater amount of solar radiation reaches the equator than reaches the polar areas primarily because of Earth's curvature. The sun's rays travel for such a long distance to Earth that they approach as parallel rays. Those rays that hit the equator strike almost perpendicularly, and thus have greater potential for heating; those hitting the poles come in at lower angles, thus less heating. To visualize this effect, imagine shining a flashlight onto water. The lower the angle of the light hitting the

Table 2.1
Length of daylight (hours:minutes) on the twenty-first of each month (2001) for various cities across New England.

City	J	F	M	A	M	J	J	A	S	O	N	D
MAINE												
Augusta	9:29	10:45	12:05	13:32	14:48	15:11	14:43	13:31	12:05	10:42	9:27	8:58
Bangor	9:29	10:44	12:04	13:31	14:46	15:10	14:41	13:29	12:04	10:40	9:25	8:56
Eastport	9:28	10:43	12:03	13:30	14:46	15:09	14:40	13:29	12:03	10:40	9:25	8:55
Ellsworth	9:29	10:44	12:04	13:31	14:47	15:10	14:41	13:30	12:04	10:41	9:26	8:56
Portland	9:32	10:48	12:08	13:35	14:51	15:14	14:46	13:34	12:08	10:45	9:30	9:01
Presque Isle	9:22	10:37	11:57	13:24	14:40	15:03	14:34	13:23	11:57	10:34	9:19	8:49
NEW HAMPSHIRE												
Berlin	9:31	10:46	12:06	13:33	14:49	15:12	14:43	13:32	12:06	10:43	9:28	8:58
Keene	9:35	10:50	12:10	13:37	14:53	15:16	14:47	13:24	12:10	10:47	9:32	9:02
Manchester	9:35	10:50	12:10	13:37	14:53	15:16	14:47	13:24	12:10	10:47	9:32	9:02
Portsmouth	9:35	10:50	12:10	13:37	14:53	15:16	14:47	13:24	12:10	10:47	9:32	9:02
VERMONT												
Brattleboro	9:35	10:51	12:11	13:38	14:53	15:17	14:48	13:37	12:11	10:47	9:32	9:03
Burlington	9:29	10:44	12:04	13:31	14:47	15:10	14:41	13:30	12:04	10:41	9:26	8:56
Rutland	9:32	10:48	12:08	13:35	14:50	15:14	14:45	13:34	12:08	10:44	9:29	9:00
St. Johnsbury	9:29	10:44	12:05	13:32	14:47	15:11	14:42	13:31	12:05	10:41	9:26	8:57

MASSACHUSETTS

Boston	9:37	10:52	12:12	13:39	14:55	15:18	14:49	13:38	12:12	10:49	9:34	9:04
Brockton	9:37	10:52	12:12	13:39	14:55	15:18	14:49	13:38	12:12	10:49	9:34	9:04
Fall River	9:38	10:53	12:13	13:40	14:56	15:19	14:50	13:39	12:13	10:50	9:35	9:05
Lawrence	9:37	10:52	12:12	13:39	14:55	15:18	14:49	13:38	12:12	10:49	9:34	9:04
Pittsfield	9:37	10:52	12:12	13:39	14:55	15:18	14:49	13:38	12:12	10:49	9:34	9:04
Springfield	9:38	10:53	12:13	13:40	14:56	15:19	14:50	13:39	12:13	10:50	9:35	9:05
Worcester	9:31	10:52	12:12	13:39	14:58	15:18	14:49	13:38	12:12	10:49	9:34	9:04

RHODE ISLAND

Providence	9:38	10:05	12:14	13:41	14:56	15:20	14:51	13:40	12:14	10:50	9:35	9:06

CONNECTICUT

Bridgeport	9:41	10:56	12:16	13:43	14:59	15:22	14:53	13:42	12:16	10:53	9:38	9:08
Hartford	9:38	10:54	12:14	13:41	14:57	15:20	14:52	13:40	12:14	10:51	9:36	9:07
New Haven	9:40	10:55	12:15	13:42	14:58	15:21	14:52	13:41	12:15	10:52	9:37	9:07
New London	9:40	10:55	12:15	13:42	14:58	15:21	14:52	13:41	12:15	10:52	9:37	9:07
Norwalk	9:40	10:55	12:15	13:42	14:58	15:21	14:52	13:41	12:15	10:52	9:37	9:07
Waterbury	9:40	10:55	12:15	13:42	14:58	15:21	14:52	13:41	12:15	10:52	9:37	9:07

Calculated from sunrise and sunset charts in Thomas (2000), *Old Farmers Almanac*, 2001. Differences from year to year are minor.

water, the greater the potential for reflecting the incoming light and the lower the amount of light per surface area.

Because energy moves from an area of greater energy to one of less energy, the excess solar energy that reaches the equator results in a net movement of warmer air toward the poles. When combined with the rotation of Earth and the resulting Coriolis effect, these air currents form the world's major circulation patterns. Those patterns loom large in relation to New England's weather and climate.

The global pattern of air flow and circulation systems can be divided into three zones in each hemisphere that migrate with the seasons (figs. 2.5 and 2.6). In the northern hemisphere, these zones include the tropics (0 to ~30°N), mid-latitudes (30 to 60°N), and polar regions (60 to 90°N). During winter, the boundary of these zones shifts southward. At the same time, the level of the tropopause—the boundary between the troposphere and the stratosphere—is lower going toward the pole, and it too shifts southward in winter (see fig. 3.3).

The entire sequence of air movement begins with the abundant heating at the equator. The air rises due to convection, moves northward, and cools in the process. As it moves northward, this cooler air sinks back to the surface at around 30°N, where it is pushed back toward the equator by the movement of cooling air behind it (fig. 2.5). This vertical motion of air is referred to as the Hadley Cell. It has a counterpart in the southern hemisphere. At the point where the air sinks, semi-permanent high pressure systems form. These high pressure systems are referred to as the subtropical highs and they are found in both of the major oceans (fig. 2.6). They vary in location and size with the seasons, becoming stronger and located farther poleward in summer, on average; they are not as strong and located farther equatorward during winter, on average. For ease in identification, these subtropical high pressure systems are referred to as the Bermuda-Azores High in the Atlantic Ocean and the Hawaiian High in the Pacific. The Bermuda-Azores High has a tremendous influence on the weather and climate of New England, whereas the Hawaiian High has only a slight influence on New England. Very often, meteorologists in the media will make reference during their forecasts to the Bermuda High, for short.

Given that high pressure systems in the northern hemisphere are characterized by clockwise, divergent flow outward from the center, the flow around the west side of the Bermuda-Azores High can extend up into New England. The air mass originating on this side of a high pressure system in the northern hemisphere is characterized by warm, moist air. Because this air originates over the tropical to sub-tropical Atlantic Ocean or the Gulf of Mexico, it is characterized as maritime tropical (mT; fig. 2.7).

On the northern limb of the subtropical highs, air flow is northward until it reaches roughly 45 to 60°N. As the overall circulation moves through the mid-latitudes, it is forced to rise again around the 45 to 60°N region, as it encoun-

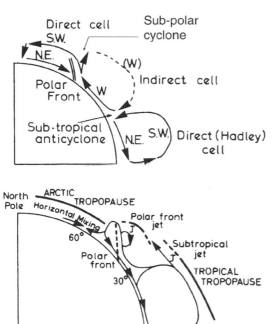

Fig. 2.5. General circulation in the northern hemisphere (*top*) showing the overall vertical and horizontal movement of air from the equator toward the poles and the resulting cells that form. The bottom figure illustrates the more specific movement of air during the winter with the location of the subtropical and polar jets (*J*) and the relative difference in height between the arctic and tropical tropopause. Both figures are shown in a cross-sectional view. The latitudinal extent of New England is 41° to 47.5°N, thus it is consistently located close to or north of the polar front and the polar jet. Modified from Barry and Chorley (1998, *Atmosphere, Weather & Climate*, 7th ed., Routledge; fig. 6.19, p. 130 [top] and fig. 6.22, p. 132 [bottom]). Used with permission from Taylor & Francis Books Ltd.

ters colder and more dense air from the polar regions. This produces an area of instability called the polar front, which is the boundary between moist tropical air and dry, continental polar air. The source of this polar air is the subsiding air associated with the high pressure that forms over the north pole. Fronts are the "lines" on Earth's surface that demarcate air masses of different characteristics, such as differences in temperature and/or moisture content. Air moving poleward from the tropics begins to rise where the formation of semi-permanent subpolar low pressure systems occurs. The subpolar low pressure system in the Atlantic is referred to as the Icelandic Low, while the analogous system in the Pacific is the Aleutian Low (fig. 2.6). Both subpolar lows have an influence on New England climate and weather. Low pressure systems in the northern hemisphere are characterized by counterclockwise, converging flow toward the center of the low pressure system. This means that New England is on the west to southwest side of the Icelandic Low and is highly influenced by the flow originating on the cooler north side of the low. This air originates over the northernmost part of the Atlantic Ocean, thus it is referred to as maritime polar air (mP; fig. 2.7). Although direct air flow from the Aleutian Low generally does not reach New England, this center of low pressure is one of the most prolific areas of storm development on Earth's surface. Consequently, most storms forming in the northern Pacific during winter move across the country and

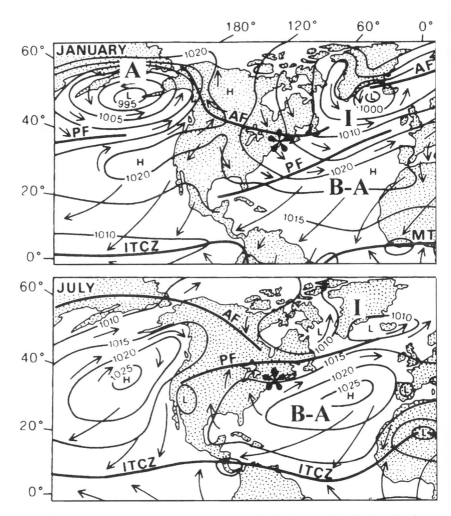

Fig. 2.6. Major circulation features in the Atlantic and Pacific Oceans of the Northern Hemisphere for January (*top*) and July (*bottom*) that have an influence on New England's weather and climate. Location of New England shown by ✳. Thin lines are lines of equal pressure at 5 millibar intervals. The general direction of air movement around these features is shown by arrows. Note the decrease in strength and extent of the Bermuda-Azores High (B-A) during winter compared to summer, the increase in strength and extent of the Icelandic Low (*I*) during winter compared to summer, and the absence of the Aleutian Low (*A*) during summer. Also note the position of the Northern Hemisphere Polar Front (*PF*) and the Arctic Front (*AF*) relative to New England in both January and July. Other global features shown here, but not discussed in the text, include the Intertropical Convergence Zone (*ICTZ*) and the Monsoon Trough (*MT*). Modified from Barry and Chorley (1998, *Atmosphere, Weather & Climate*, 7th ed., Routledge; fig. 7.19, p. 169). Used with permission from Taylor & Francis Books Ltd.

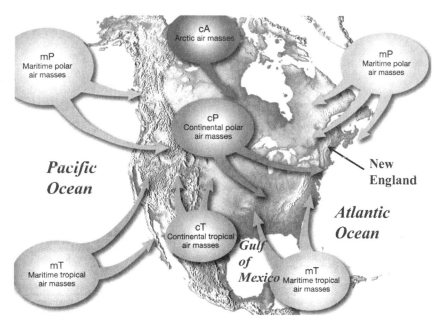

Fig. 2.7. Major air masses that affect North America including New England. Courtesy of WARD'S Natural Science Establishment, Inc.

eventually have an impact on New England. The Aleutian Low does not exist in this same form or location during the summer. Figure 2.8 summarizes the equator-to-pole circulation in the northern hemisphere and the change in the location of the polar front with the seasons as well as the nature of the air masses that affect New England with changes in the source area relative to New England.

Westerlies

So far, we have been describing primarily the vertical motion of air movement through the northern hemisphere as brought on by the unequal distribution of solar radiation and heating of Earth's surface. Important zones of horizontal air flow (called advection) across the globe coincide with the three vertical zones. The major zone of horizontal air flow that controls New England's weather and climate is the mid-latitude westerlies (airflow from west to east) that occur both at the surface and at upper levels (that is, at altitudes above 25,000 feet). Surface westerlies form via air flow around the northern side of the subtropical highs in conjunction with Coriolis force. Flow around the subpolar lows also contributes to the formation of the surface westerlies.

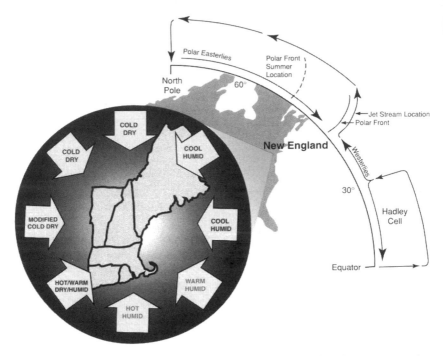

Fig. 2.8. Summary of weather conditions associated with air masses of varying source areas that reach New England and their relationship to general northern hemisphere circulation. Used with permission from *New England's Changing Climate, Weather and Air Quality*, Climate Change Research Center, University of New Hampshire, Durham, New Hampshire 03824, © 1998, Climate Change Research Center.

The important contribution that the upper-level westerlies make to New England's climate is two-fold. First, the westerlies are responsible for bringing air parcels across the country and because the major flow is west to east, New England actually has a climate with a high degree of continentality. That is, despite the proximity of the ocean to New England, most weather and climatic conditions are dominated by conditions over the landmass to the west of New England. Secondly, the westerlies influence location of storm tracks as a function of the shape of the air-flow pattern. Two main patterns exist for the westerlies. One pattern is referred to as zonal flow, whereby the main airstreams are west to east with not very much of a north-south component to them (fig. 2.9). When the westerlies are in this form, air originating from the Pacific Ocean can reach New England, although it is modified. This produces mild temperatures in the region. On the other hand, the westerlies can contain a series of waves or north-south components, referred to as a series of troughs—with their open side to the north—and ridges—with open side of the wave to the south. This scenario is referred to as meridional flow. The nature of the air reaching New England is a function of what part of the wave pattern is near New England, and of the severity of the wave pattern. Meridional circulation leads to much greater changeability in New England weather patterns.

The part of the upper-air westerly flow that is most commonly used to describe its pattern is the jet stream. This is the area within the westerlies where the air flow is the fastest. The position and strength of the jet stream greatly influences the direction and severity of storms, not to mention how quickly one can travel across the country in a plane. Furthermore, the jet stream is basically the upper-level manifestation of the surface position of the polar front (fig. 2.5). This is not to be confused with the subtropical jet that forms as the upper-level manifestation of the sinking arm of the Hadley Cell. The position of these jet streams thus can dictate whether New England is on the cool side of the polar front or the warm side. We will discuss the variability in the seasonal position of the polar front in more detail as we discuss specifics about the seasons.

Another front in the northern hemisphere occasionally influences New England's climate, but only during winter. This is the Arctic front, which marks the boundary between the frigid air over the snow cover of the northern hemisphere during winter and the cold-to-cool air over snow-free ground (fig. 2.6). Air to the north of the Arctic front basically is coming right over the north pole toward New England.

Air masses that originate over the snow and ice cover of Canada and the northernmost polar regions are very cold and dry. New England also feels these

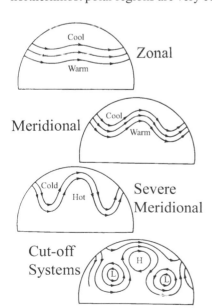

Fig. 2.9. Various shapes taken by the westerlies in the northern hemisphere ranging from zonal (top) to various degrees of meridional flow (middle) to times when the westerlies are disjointed (bottom). The latter shape results in cut-off low pressure systems and blocking high pressure systems. General temperatures on either side of the main jet stream are shown for relative comparisons. Actual temperature regimes would be a function of season. The example of the cut-off systems shown is typically referred to as an Omega Block because the overall air flow is similar to the Greek letter Ω. Modified from Barry and Chorley (1998, *Atmosphere, Weather & Climate,* 7th ed., Routledge; fig. 6.25, p. 136). Used with permission from Taylor & Francis Books Ltd.

fresh, crisp air masses from Canada during summer. As a result, New England's weather is highly influenced by continental, polar air masses (cP; fig. 2.7). During winter, frigid air masses coming from directly over the pole are often referred to as Arctic air masses (cA).

Occasionally, we can be affected by dry, hot air coming across the United States toward New England that originated in the Southwest or Texas. These occurrences are very rare, but we occasionally do feel the impact of these continental tropical air masses (cT; fig. 2.7), although they are somewhat modified as they move across the country. In general, the source of continental air masses reaching New England is a function of the position of high pressure systems moving across the country.

High and Low Pressure Systems

As the final components of the weather systems that affect New England, we now give several examples of the typical position of highs, lows, their associated fronts, and the resulting conditions across the region. These are the items one would see on daily weather maps. Although we are characterizing these conditions by the dominant positioning of individual pressure systems, it is important to remember that individual systems work together to produce the conditions observed across New England.

High pressure to the west and northwest. This scenario is the classic example of the situation that exists following the passage of a low pressure system and trailing cold front across New England (fig. 2.10). The position of the high to the west of New England results in fair conditions with a flow of cool or cold dry Canadian air, often on brisk northwest breezes. The strength of the winds associated with this scenario depends on how strong the low pressure system was that just moved out of New England. A strong low in combination with a strong high produces a large difference in central pressures, and thus a steep pressure gradient between the two systems and strong winds.

High pressure to the east and southeast. As the high pressure system in the scenario mentioned above moves to the east or southeast, the air flow to New England is on the west side around the central part of the high (fig. 2.11). This flow brings warm or hot humid air into New England. This scenario is a classic summertime phenomena that produces those days with the three H's: hazy, hot, and humid. Often the high may stay in that position for a long enough time that the air over New England becomes very stagnant and problems with pollution (e.g., ozone levels) can exist.

High pressure to the northeast. This situation brings unexpected conditions to New England, since high pressure systems are usually associated with fair

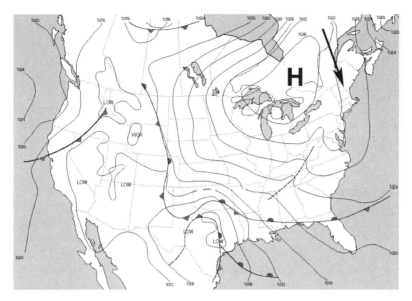

Fig. 2.10. Classic example of a high pressure system located to the west-northwest of New England. Airflow to New England is generally from the north to northwest as shown by the arrow. This map is a general version of synoptic conditions for 16 March 1998.

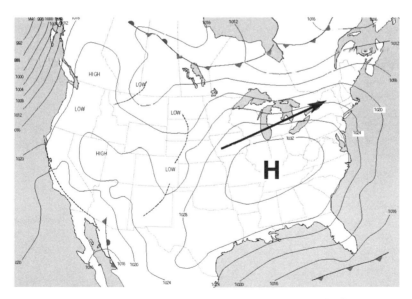

Fig. 2.11. Classic example of a high pressure system located to the south of New England. As the high drifts more to the east from the position shown here, airflow to New England changes from the southwesterly flow depicted here (shown by the arrow) to more of a south-southwesterly flow. This map is a general version of synoptic conditions for 23 October 1998.

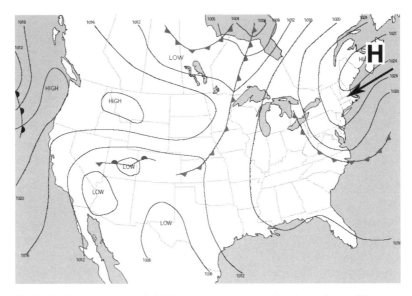

Fig. 2.12. Classic example of a high pressure system located to the northeast of New England. Airflow to New England is generally from the northeast, as shown by the arrow. This situation may produce a backdoor cold front, as shown here and discussed in chapter 13 (see figs. 13.2 and 13.3). This map is a general version of synoptic conditions for 10 June 1999.

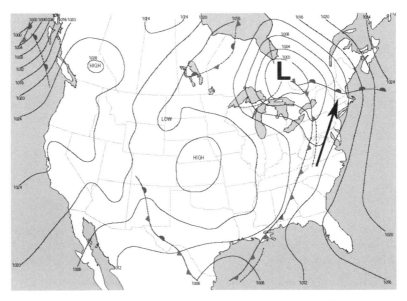

Fig. 2.13. Classic example of a low pressure system located to the west of New England. Airflow to New England is generally from the south-southwest as shown by the arrow. This map is a general version of synoptic conditions for 17 April 1998.

weather (fig. 2.12). The flow around the southeast side of the high pressure system brings in cool, moist air off the Atlantic Ocean. This can produce low-level clouds, fog, light rain, or ocean-effect snow.

Low pressure to the west. The classic low pressure system is associated with a warm front to the east of the low and a trailing cold front extending to the southwest of the low (fig. 2.13). When this system moves northeastward through the Saint Lawrence River valley, west of New England, more precipitation will fall along the western side of the region closer to the low, as well as along the trailing cold front as it sweeps across the area.

Low pressure to the east. In this situation, the counterclockwise flow around the low pressure system will bring an air mass off the ocean, and with it the potential for copious amounts of precipitation (fig. 2.14). This classic coastal storm or "nor'easter" is an integral part of New England's weather and climate. We discuss nor'easters in much more detail in chapter 15.

Summary

Spatial and temporal changes in weather and climate across New England begin with day-to-day changes. These diurnal changes are controlled by changes

Fig. 2.14. Classic example of a low pressure system located to the east of New England. Airflow to New England is generally from the southeast to the northeast (as shown by the arrow) depending on storm track and where the center of the low pressure is located at any one time. This map is a general version of synoptic conditions for 2 May 1998.

in solar input to the surface of Earth and the ability of the land and ocean surface to retain and lose heat. Changes in the angle of the sun over the horizon vary with the seasons, further influencing the amount of heating that occurs at the surface. The higher angle of the sun's rays and longer period of daylight in the summer produces warmer temperatures compared to the lower temperatures encountered during winter with the lower angle of the sun and shorter daylight hours.

The distribution of solar energy across the globe and the spin of Earth—the Coriolis effect—also control New England weather and climate through the development of various circulation systems around the world. The systems that have the greatest impact on New England climate are the subtropical Bermuda-Azores High and subpolar Icelandic Low, both located in the North Atlantic Ocean. The clockwise circulation around the Bermuda High (for short) brings warm, humid air to New England, whereas the counterclockwise flow around the Icelandic Low brings cool, humid air to the region. These sytems, together with the rotation of Earth, produce the westerlies that dominate airflow across the mid-latitudes, including New England. Zonal flow, that is, a relatively flat flow across the country, produces mild conditions in New England, whereas meridional flow of the westerlies—the presence of troughs and ridges—brings together contrasting air masses and conditions that perpetuate storm development. The positions of day-to-day high and low pressure systems and their associated frontal systems control our weather and climate (temperature and precipitation regimes) through variability in wind flow and air mass type.

Chapter 3

Year-to-Year Changes

In Maine everyone pays attention to the weather. —J. S. BORTHWICK

One interesting aspect of New England weather and climate is the great variability that occurs from year to year. New Englanders are constantly wondering whether the upcoming winter will provide enough snow for excellent skiing conditions. Individuals in the tourist industry are always eager to know how hot and dry or cold and wet the summer will be, because cool, wet summers may very well deter beach and lake goers. Major differences in circulation patterns from year to year during the transition seasons, spring and fall, result in many weather surprises. A heavy April snowstorm may occur in one year while on the same day in the following year, temperatures may reach into the 90s. Several factors contribute greatly to this interannual variability as well as causing changes from decade to decade. For example, the mid-1960s to mid-1970s were certainly cooler than the 1980s and 1990s. We now briefly describe the factors that force climate changes on an annual basis, and give some examples of years when these mechanisms played a major role in dictating prevalent conditions.

El Niño and La Niña

During the winter of 1997/1998, El Niño was one of the most influential factors affecting not only New England climate, but global climate as well. We will not discuss the media blitz that made El Niño the scapegoat for many natural events during that winter. El Niño and its opposite counterpart, La Niña, are the two end members of changes that take place in the ocean and atmosphere of the tropical Pacific Ocean. These changes in sea level pressure and sea surface temperature are part of a larger system of fluctuating conditions in the southern Pacific Ocean known as the Southern Oscillation. The entire system is referred to as the El Niño-Southern Oscillation (ENSO).

An El Niño event occurs when the warm ocean waters around northern Aus-

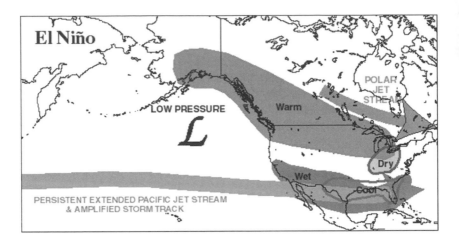

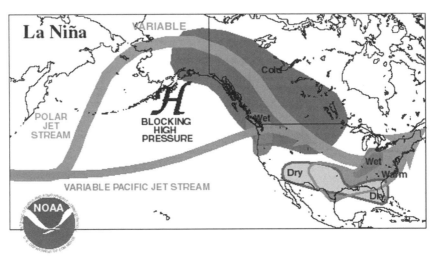

Fig. 3.1. Common position and shape of the northern hemisphere jet streams during the winter of moderate to strong El Niño (*top*) and La Niña (*bottom*) events. Anomalous climatic conditions across the country are also shown. Note that the polar jet is usually north of New England during moderate to strong El Niño events and may often be over New England during moderate to strong La Niña events. Both cases would produce warm winter temperatures in New England.
Maps from the Climate Prediction Center, National Oceanic and Atmospheric Administration/Department of Commerce, web page at http://www.cpc.ncep.noaa.gov/products/analysis_monitoring/ensocycle/nawinter.html. This website URL, and all others that appear in this book, were up and running at the time of publication.

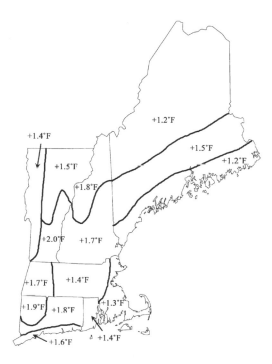

Fig. 3.2. Average amount of winter temperature increase above average winter temperature during El Niño years for the climatic zones defined in each New England state. The months used to determine averages are December to February for Maine, New Hampshire, and Vermont and November to December for Massachusetts, Rhode Island, and Connecticut. Years used in the analysis were 1940 to 1941, 1957 to 1958, 1965 to 1966, 1972 to 1973, 1982 to 1983, 1986 to 1987, 1991 to 1992, and 1994 to 1995. Climatic data from each zone were used to produce the state-wide average temperature and precipitation figures in chapter 19. Modified from the Climate Prediction Center, National Oceanic and Atmospheric Administration/Department of Commerce, web page at http://www.cpc.ncep.noaa.gov.

tralia and Tahiti migrate eastward with a weakening of the easterly trade winds. As a result, a warm pool of water forms in the tropical Pacific Ocean off coastal South America. These conditions bring stormy weather to portions of the west coasts of North America such as southern California, as well as to the western coast of Central and South America. El Niño also causes a shift in the jet stream across the United States (fig. 3.1).

The primary effect on New England of this shift in circulation is an increase in winter temperatures in the range of 1 to 2°F (fig. 3.2). However, the exceptionally strong 1997/1998 El Niño produced some of the warmest winter temperatures ever recorded, with Connecticut averaging 5.7°F warmer than mean winter temperatures and New Hampshire averaging 5.9°F warmer than the mean. Southern New England and New Hampshire appear to have warmer winter temperatures with greater consistency during an El Niño year than do Maine and Vermont. It is natural to assume that these warmer winter temperatures may result in a lessening of snowfall totals. However, this scenario does not necessarily hold true, as both Portland and Boston receive only a few inches less snow in moderate to strong El Niño winters than in neutral years (that is, years when neither an El Niño or La Niña is in existence). Precipitation during years under the influence of an El Niño does not change significantly, except that April through June may be drier than normal, particularly in northern New England.

Prior to the 1997/1998 El Niño, the 1982/1983 El Niño event was the most powerful of the past few decades. However, El Niño events are variable in their overall magnitude, so that the impact on New England is not always as extreme as that in 1997/1998 (fig. 3.2). Over the last few decades, El Niño events have occurred every three to seven years, on average, and persisted for a year or two. However, an El Niño persisted over the entire four-year period from 1991 to 1994. Past changes in global climate are thought to produce different cycles than the three to seven years that has been occurring over the past few decades.

During a La Niña event, the easterly trade winds in the Pacific Ocean are stronger and the pool of warm water remains in the western Pacific. Also, stronger ocean currents deliver colder than normal waters to the equatorial Pacific near South America. In this case, storminess is much reduced along the western coast of the Americas, with a different shift in overall circulation patterns across North America compared to those created by an El Niño event (fig. 3.1). However, the resulting climatic conditions in the United States, and particularly in New England, have not been as well documented for a La Niña event as for El Niño events, possibly because the position of the jet stream seems to be more variable for La Niña events than for El Niño events. For instance, La Niña conditions were in existence during the winter of 1998/1999 and climatic conditions across New England were not opposite to that of an El Niño year. In fact, the winter of 1998/1999 was a very low snow year much like that of the El Niño–dominated winter of 1997/1998. Moreover, temperatures during the 1998/1999 winter were warmer than average in some areas, conditions normally characteristic of an El Niño.

Nevertheless, some general trends in temperature and precipitation across New England have been observed during La Niña years, with northern and southern New England often behaving differently. For example, southern New England and northern coastal areas of New England appear to be generally warmer than average during the latest part of winter (January to March). Northernmost Maine may be a bit cooler. Another general trend is that almost all of northern New England appears to have cooler springs, while southern New England is likely to continue with warmth during the spring of a La Niña year. Both northern and southern New England may have slightly cooler summers. Overall, precipitation amounts may be above average, particularly in late winter and late summer, across the region. Surprisingly, La Niña years appear to be slightly less snowy in Portland than both neutral and El Niño years, but only by a few inches. Snowfall amounts for La Niña years in Boston are between neutral and El Niño years. Unlike an El Niño event, the response of New England to La Niña conditions is much more variable.

The cycle for the timing of La Niña events is similar to that for El Niño given that the La Niño phase of the oscillation is the other extreme. However, in some years, neither a well-defined El Niño nor a La Niña exists. Other factors may play a greater role in dictating New England's weather and climate for that particular year.

Volcanic Eruptions

On 6 June 1816, several inches of snow were recorded as far south in New England as western Massachusetts (Stommel and Stommel, 1983). This was just the beginning of three days of sporadic snowfall. In addition, several occurrences of heavy frost were noted in all three summer months (June to August) of that year. Daily maximum temperatures were often in the 60s, and even in the 50s in July. These phenomena led to 1816 becoming known as the "Year without a Summer." This extremely cold summer can be directly attributed to the very large, explosive eruption of Tambora volcano, located on the island of Sumbawa, Indonesia, the year before (10 April 1815). Sumbawa is located halfway around the world from New England. This is an extreme case of volcanic impact on New England's climate; the Tambora eruption is the largest known eruption of historical time. Nevertheless, eruptions that are considered only moderate in magnitude by geological standards can produce cooler climatic conditions in New England and around the world.

When a volcano erupts explosively, it spews out two main products. One component of the eruption is the mineral or silicate matter that produces widespread ash deposits or lava depending on the type of eruption. The second component emitted is gas, which includes such elements as sulfur, carbon, chloride, and fluoride, as well as water vapor. Once these gases are injected into the atmosphere, they react with other compounds (such as water), possibly forming acid particles or aerosols such as sulfuric acid and hydrochloric acid. The presence of both the ash and acidic aerosols in the lower parts of Earth's atmosphere block incoming solar radiation, thus cooling Earth's surface. However, ash particles are much too heavy to stay aloft for a long period of time. Consequently, they have very little impact on climatic conditions anywhere in the world except for a few days close to the location of the eruption. On the other hand, the acidic droplets are lighter and can remain aloft in large quantities for three to four years. These droplets both absorb and reflect incoming solar radiation, preventing it from reaching Earth's surface. The type of acid that is primarily responsible for cooling Earth's climate is sulfuric acid, because many of the other acids (including hydrochloric acid) are much more soluble and do not stay aloft for a long period of time. These other acids wash out of the atmosphere very quickly. Acids that remain in the lowest part of the atmosphere, the troposphere, where our weather takes place will be washed out by precipitation within a week to ten days (fig. 3.3). For an adequate amount of aerosols to remain aloft for three to four years, the height of the eruption column must penetrate the tropopause, thus reaching the stratosphere, where there is no precipitation to wash out the sulfuric acid droplets. The acid eventually falls by gravity into the troposphere, where it will be cleansed from the atmosphere by precipitation.

In summary, for a volcanic eruption to have an impact on New England's climate, it must be explosive enough to inject material into the stratosphere and it must eject sulfur-rich gases that eventually convert to sulfuric acid. Eruptions that impact New England must be located either in the equatorial regions (about 20°N to 20°S latitude) or in the northern hemisphere. The stratospheric cloud produced by an equatorial eruption will be spread into each hemisphere, but too little volcanic material crosses the equator for a mid-latitude southern hemisphere eruption to have an impact on New England's climate. A volcanic eruption that meets these criteria will cool the earth's climate by 0.5 to 0.7°F during the first year or two following the eruption, with less cooling over the next year or two. Aerosols from the eruption that still remain aloft after three or four years usually have a negligible impact on climate. We show the evidence for how specific volcanic eruptions during the twentieth century affected New England's climate in chapter 19.

Greenhouse Gases

The accumulation of greenhouse gases in the atmosphere does not produce an immediate impact on New England's climate like that of an El Niño event or a volcanic eruption. However, the overall warming in temperature in parts of New England almost year after year through the 1980s and 1990s is possibly related to increasing greenhouse gases (see chapter 19 for details). One year that deviated from this trend was 1992, the cool year resulting from the eruption of Mt. Pinatubo. It is noteworthy that some areas of New England have not experienced an exceptionally distinct increase in temperature over the last two decades, although a discrepancy in the overall trend is not unexpected. Many climate-forcing components may provide an average warming or cooling over a region and especially over hemispheres or over the entire globe, while at the same time other areas may experience completely opposite climatic conditions. For example, a specific area within a region such as New England may be experiencing overall cooler conditions or no real changes over a certain length of time, while the average for all of New England over that same length of time is one of warming.

The gases with the most powerful greenhouse effect are carbon dioxide, methane, and water vapor. Carbon dioxide (CO_2) from the burning of fossil fuels is the most publicized of the greenhouse gases, but water vapor is a far more effective greenhouse gas. Gases such as CO_2 absorb and re-admit long-wave

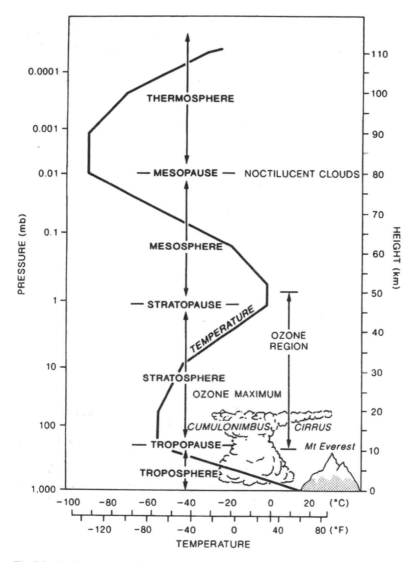

Fig. 3.3. Vertical structure of the atmosphere showing the change in temperature with height that defines the various layers of the atmosphere. The tropopause, which marks the boundary between the troposphere—where our weather takes place—and the stratosphere, is at an average elevation of about 15 km (9.3 miles) in the tropics and about 10 km (6.2 miles) in the polar regions (see fig. 2.5, bottom). The elevation of the tropopause is lower during the winter than in the summer in each hemisphere. Modified from Barry and Chorley (1998, *Atmosphere, Weather & Climate*, 7th ed., Routledge; fig. 1.15, p. 16). Used with permission from Taylor & Francis Books Ltd.

radiation back to Earth's surface. Long-wave radiation is absorbed solar radiation that is emitted by Earth's surface. Methane may be an especially important gas should warming continue, as there is probably much methane locked up in permafrost regions of the high latitudes of the northern hemisphere. Continued warming could thaw some of these areas, thus increasing methane release into the atmosphere and at the same time continuing the increase in global temperatures and possibly New England's temperatures.

Solar Activity

Just like greenhouse gases in the atmosphere, variability in the intensity of solar radiation produced by the sun causes fluctuations in Earth's climate over decadal time frames, as opposed to abrupt changes from year to year. Output from the sun is not constant, as many had believed even at the beginning of the twentieth century. In addition, the amount of sunspot activity affects the intensity of solar output. The intensity of sunspot activity fluctuates on eleven- and twenty-two-year cycles; records of climatic change clearly show these fluctuations. For example, temperatures may gradually rise to a peak over roughly a decade, then fall back over the next decade. Temperatures may then rise again over another decade, thus following this eleven-year cycle.

Adding further to the eleven-year solar influence on New England's climate, overall solar activity has been either higher or lower than average during periods of several decades over the last few centuries. The time periods of consistent solar activity have been identified and named, such as the Maunder Minimum (1645–1715), Spörer Minimum (1416–1534), and Wolf Minimum (1282–1342), periods when sunspot activity was non-existent (as in the Maunder Minimum) or was lower than average. Temperatures were lower than the preceding and subsequent decades surrounding these time frames. Periods when solar activity was higher than average would result in warmer than average temperatures around New England. It is obvious from these records that changes in solar output have been a very prevalent component of New England's climate during several decades just prior to and since European inhabitation of New England.

The Random Component

One final component in the climate system can have a huge impact on the annual variability in climate. The factor in question is the "noise" inherent in the climate system, that is, the year-to-year variability that cannot be accounted for by any distinct forcing factor. For instance, circulation patterns may be such in a particular year that much of the airflow reaching New England comes from the northwest, thereby producing cooler than average temperatures for a significant part of that year. Such patterns occurred during several winters in the 1800s (Ludlum, 1966; 1968). In several cases, it was recorded that during the entire month of January or December into January, winds were either from the northeast or northwest, resulting in cold temperatures and a high number of snowstorms. Interestingly, the remainder of some of those winters were rather warm with a lack of snow. Records indicate that total snowfall was probably close to average in those particular winters, but much of that snowfall was concentrated into a very short period of time. It is hard to say why specific circulation patterns may prevail over extensive periods, but they can lead to lengthy—weeks or even months—with very similar climatic conditions. The cool summer of 2000 was a direct result of the prevalance of northwest flow of the jet stream over New England. Such a distinct time period when specific conditions are prevalent may play a significant role in determining average conditions for that year. All natural systems have this noise component within them. In the case of the climate system, the noise component, when added to other forcing factors, produces the year-to-year and decade-to-decade variability that characterizes the climate system of New England.

Summary

Annual variability in New England's climate results from a combination of many different factors. Very warm winters are known to have been highly influenced by El Niño events. Particularly strong El Niño events, as occurred during the winters of 1997/1998 and 1982/1983, can create exceptionally strange winters. Volcanic eruptions are known to cause cool summers in New England, as has been so vividly detailed in the many accounts from 1816, the "Year without a Summer." In 1992, in the middle of the recent warming trend, New England experienced a very cool summer, which followed the 1991 eruption of Mount Pinatubo in the Philippines. Beach and lake tourism was also very low that summer—who wants to use blankets while watching the Fourth-of-July fireworks!—showing the impact that these natural events can have on New England society.

The point that should be taken from this discussion is that many forcing factors can cause annual to decadal variability in New England's climate system. These factors can be working at the same time and, in fact, opposing factors may end up producing an "average" year. It is known that the cooling associated with a volcanic eruption may be much less than expected if the eruption coincided with warming from an El Niño event. This was exactly the case in 1982 to 1983. The Mexican El Chichón eruption had the potential to cool climate almost as much as the Pinatubo eruption, but temperatures in the summer of 1983 were not as cool as expected because of the very strong El Niño in 1982 to 1983. All of these complex relationships definitely contribute to the great year-to-year variability in New England's weather and climate.

Chapter 4

Changes over Long Time Periods

*I reverently believe that the Maker who made us all makes everything
in New England but the weather.* —MARK TWAIN

Fluctuations in climate also take place over time scales that extend well
beyond a human lifetime. For example, Earth's climate over the past
two million years has oscillated between periods called glacials, when
large ice sheets dominated the continents of the northern hemisphere, and
interglacials, which are much like the modern-day climate. Glacial periods
probably experienced average global temperature about 8 to 10°F colder than
current values, and persisted for much longer durations (about 90,000 years
each) than interglacials. In contrast, interglacial periods experienced temperatures near modern levels and persisted for about 10,000 years.

These major climatic fluctuations are related to changes in Earth's orbit,
which then changes the distribution of solar radiation across Earth's surface.
These fluctuations are part of the Milankovitch theory of climatic change.
These cycles include changes in the shape of Earth's orbit around the sun
(a 95,800-year mean cycle between an elliptical and spherical path), fluctuations in the tilt of Earth's axis of rotation (a 41,000-year mean cycle between
21.8° and 24.4°), and variations in the season of year when Earth is closest to
the sun (a 21,700-year mean cycle). Collectively, these cycles lead to varying
intensities of the glacials and interglacials, but, in general, Earth experiences a
100,000-year oscillation, with glacial periods lasting 90,000 years and interglacial periods lasting 10,000 years. It is interesting that it takes a long time
to build up large ice sheets, but only a short time to melt them away. While
the Milankovitch theory largely explains the cyclical climate of the past two
million years and beyond, other important mechanisms in the climate system
contribute to the shifts between glacial and interglacial conditions. For instance, changes in oceanic circulation play a role in these processes. Even
more interesting is the fact that shifts from near-glacial to near-interglacial
conditions occurred in less than a decade, particularly at times when ice sheets
existed on the northern hemisphere continents. Clearly, the climate system is

Fig. 4.1. Tuckerman Ravine, in the White Mountains of New Hampshire, is a classic cirque. Photograph from http://www.tuckerman.org/photos/tucks/4–7–01.jpg. Photo provided by the United States Forest Service.

exceptionally dynamic, even more so than was thought just a few decades ago. The analyses and results available from the Greenland Ice Sheet ice cores collected in the 1990s, as well as from other ice cores collected from Antarctica, have provided extensive evidence of the factors that govern Earth's climate and its frequent and rapid fluctuations (Alley, 2000; Mayewski and White, 2002).

The impact of these glacials and interglacials on the modern New England landscape is profound, primarily through the erosional and depositional processes associated with glaciers. As conditions cooled from an interglacial to a glacial, mountain glaciers probably were the first to form on the New England landscape, most likely prior to the arrival of the massive continental ice sheets. Alpine glaciers were limited to the mountainous regions of northern New England, such as the highest elevations of the White Mountains. As glacial ice formed near the crests of these mountains, processes associated with ice flow, as well as processes associated with freezing and thawing of water in rocks at the head of the valley walls, would remove pieces of rocks from the mountains to form deep ampitheater-like basins called cirques. Cirques are concentrated near the summit of Mount Washington and they include Tuckerman Ravine and Huntington Ravine (fig. 4.1). Cirques are also found on Mount Katahdin

Fig. 4.2. The traditional spring skiing on the cirque headwall of Tuckerman Ravine, White Mountains, New Hampshire. Photograph from http://www.tuckerman.org/photos/tucks/4–28–01_ lunch_rocks.jpg. Photo provided by the United States Forest Service.

in Maine, in the Chimney Pond area. Today, Tuckerman Ravine is a widely popular setting for hiking and skiing, and any serious New England downhill skiier must make the pilgrimage and ski "Tucks" (fig. 4.2).

With increasing glacial conditions, the region eventually fell under the influence of larger-scale continental ice sheets. The continental-based glaciers that originated in Canada began to increase in size with flow southward toward the north-central and northeastern United States. On several occasions, New England, in its entirety, was held in the grips of this glacial ice. The ice was so thick on occasion that lofty Mount Washington was encased below the surface of the glacier as recently as 15,000 to 20,000 years ago. At the same time, ice filled every stream valley in New England, reshaping them from V-shaped to U-shaped (in cross-sectional view). Several examples of these glacially eroded troughs include Franconia Notch, Dixville Notch, Crawford Notch, and the Great Gulf in New Hampshire (fig. 4.3).

Several other aspects of the New England landscape provide evidence of the erosive powers of these large ice sheets. Processes associated with moving ice removed material from the bedrock, in some cases forming large depressions that became the major lakes in New England. These include Lake

Fig. 4.3. Glacially carved U-shaped valley of Crawford Notch in the White Mountains of New Hampshire. Used with permission from David Metsky as at http://www.cs.dartmouth.edu/whites/ crawford.html.

Winnipesaukee, Squam Lake, and Sebago Lake in northern New England. Rocks that became imbedded in the bottom of the glacier scratched the rocks over which the ice passed, leaving distinct scratches in the bedrock referred to as striations (fig. 4.4). Striations are visible on exposed rock outcrops at numerous locations in New England, including the summit of Mount Washington. Furthermore, rocks carried long distances from their source are often found on top of bedrock of a completely different composition. These rocks, referred to as erratics, have been found on top of Mount Washington and on Cadillac Mountain in Acadia National Park and on the outer part of Cape Cod (fig. 4.5). The presence of both striations and erratics on mountain tops across New England provide the evidence needed to conclude that the entire region was covered by these large ice sheets. Striations and erratics also give scientists clues as to the direction of the ice flow, as does the bedrock feature referred to as a roche moutonnée, or "sheepback," formed by the removal of material on the downflow side of the rock (fig. 4.6). Finally, the erosive power of these ice sheets removed much of the soil in New England. Soil layers today are mostly shallow and agricultural practices are hindered by these poorly developed soils.

Unlike features formed by glacial erosion, features formed by the deposition of sediment carried by the ice are prevalent throughout New England. One type of deposit is a small hill that may have a distinctive "teardrop" shape. This feature is termed a drumlin. Drumlins form when ice conditions at the base of the glacier and the sediment burden in the lowermost layers of ice reach a

Fig. 4.4. Glacially formed striations on a rock outcrop in northwest Massachusetts. Ice moved from right to left, as indicated by the knife in the center of the picture, which also provides scale. Photo by Greg Zielinski.

Fig. 4.5. Doane Rock, Eastham, Massachusetts, the largest glacial erratic on Cape Cod. Photo by Greg Zielinski.

Fig. 4.6. Roche moutonnée located in the Connecticut River valley of western Massachusetts. Ice flow was from right to left with the "plucked" section of the rock outcrop located on the down-ice side of the rock. Photo by Greg Zielinski.

Fig. 4.7. Drumlin located in Madbury, New Hampshire. Ice flow was from right to left in this picture. In this case, the shape of the drumlin is opposite to the classic form of a steep slope on the up-ice side and a gentle slope on the down-ice side. Photo by Greg Zielinski.

certain stage that allows the deposition of the sediments. These sediments are then streamlined in the direction of ice flow by the moving ice. Often drumlins are lain down in large clusters called "swarms." Swarms of drumlins are prevalent in several areas of southern New Hampshire, Vermont, and Massachusetts (fig. 4.7). Today, drumlins are important for several reasons, such as providing a good source of material for construction purposes. Drumlins provided high ground during the initial colonization of New England by Europeans, that is, at a time when engineering was in its infancy and flooding was more problematic. Drumlins also help scientists reconstruct the direction of the ice flow during their deposition. These features may have a steep side that faces the upstream direction of ice flow, and a gently tapering slope to form the top of the teardrop, which faces in the downstream direction, although they do have many different shapes. Perhaps the most famous drumlins in New England are Bunker, Breeds, and Beacon hills in Boston and the hills in Boston Harbor (fig. 4.8).

At the maximum extent of glaciation, the edge of the ice sheets never made it far past the current shores of southern New England. Typically, a prominent feature called an end or terminal moraine forms by the accumulation of sediment at the margin of the farthest advance of the glacier. The terminal moraine from the last glaciation forms parts of the coastline that we know today as Block Island, Martha's Vineyard, Nantucket, and Long Island, New York. As the location of the snout of these glaciers retreated northward under warmer climate conditions, they often paused for a period of time at one location and

Fig. 4.8. Schematic of the battle of Bunker Hill, actually fought on the top of Breeds Hill, a drumlin. Because drumlins were high spots on the landscape, they became strategic sites during the Revolutionary War. Photo from the Discovery Channel home page. From WORLD BOOK ONLINE © 2002, World Book, Inc. By permission of the publisher, www.worldbook.com.

Fig. 4.9. Recessional moraine in south central Maine. In this particular case, the edge of the ice was in the ocean, thus this feature is a submarine moraine or DeGeer moraine. Photo by Greg Zielinski.

Fig. 4.10. Walden Pond, a famous kettle hole in New England. Used with permission from Nanosoft, Inc. of Massachusetts, as seen at http://www.nanosft.com/walden/october/trees1.html.

left behind other ridges. The resulting features are called recessional moraines (fig. 4.9). Today, these relic features include portions of the inner part of Cape Cod and Long Island. Since these features consist of deposits from glaciers, they are made up of unconsolidated rocks, sand, and silt, but they do not have the strength of the bedrock that constitutes the heartland of New England. As a result of the high wave energy in these stormy coastal environments, these features are highly vulnerable to erosion (see fig. 9.1). Some scientists estimate that Nantucket will disappear completely from erosional processes within the next five hundred years. While this may sound like a long time, it is barely a heartbeat over geologic time.

With the continued prevalance of warmer climatic conditions, the edge of the continental glacier continued to retreat northward into Canada. During this melting process, huge streams flowed off the ice, leaving large sand and gravel deposits in the valleys of New England. These deposits provide groundwater for many water supply systems throughout the region. The retreating ice also left behind large blocks of ice that, upon melting, often formed small depressions or kettle holes on the surface of these sand and gravel deposits. Often these depressions hold water, such as Walden Pond in Massachusetts (fig. 4.10). However, if the depth to the water table is greater than the depth of the kettle hole, it will remain dry, such as in Spruce Hole, near Durham, New Hampshire.

Summary

The location of New England relative to the advance and retreat of former continental and alpine glaciers has played an indelible role in shaping the landscape of the region. Many of the modern tourist features were carved out by glaciers, including Tuckerman Ravine and Franconia Notch. In addition, glacial deposits from these glaciers have been important to the history of New England, as in the case of Breeds and Bunker hills, as well as for recreational purposes, as in Nantucket and parts of Cape Cod. Even at time scales of thousands and millions of years, the climate of New England is highly variable and the impact of its most recent changes still remains.

Diversity of New England's Weather and Climate From the Mountains to the Beaches

There is a sumptuous variety about the New England weather
that compels the stranger's admiration—and regret.

—MARK TWAIN

In addition to the great changes from year to year, as well as over longer time periods, New England displays great diversity of weather and climate across the region. In short, this great spatial variability is a function of the fact that one can go from the mountains to the beaches over a very short distance. With this geographical fact, in combination with New England's location with respect to major circulation patterns of the northern hemisphere, weather and climatic conditions are as diverse or are more diverse than anywhere in the country over such short distances. This is especially true for the winter climate.

The perfect example of how spatial variability across the region can play havoc on New England weather is the forecast given for an approaching winter storm. In this part of the book, as well as in chapter 15 on nor'easters, we describe how these storms can bring all snow to mixed precipitation to all rain over very short distances within the region. Many factors have a bearing on the location of regional weather boundaries. In fact, these same factors play key roles in dictating many other aspects of New England's weather and climate.

In the two chapters of this part of the book, we present two groups of primary causes for the spatial diversity in New England's climate. One group, which we discuss in chapter 5, includes those that essentially remain permanent in one's lifetime, such as the influence of the mountains and the ocean currents. The second group, presented in chapter 6, includes factors that always exist, but that can change in position and intensity from year to year. The main components of these non-permanent climatic factors are storm tracks and variability in other northern hemisphere circulation systems.

Chapter 5

Permanent Causes

As to variety, why, he confessed that he got hundreds of kinds of weather
that he had never heard of before. —MARK TWAIN

Several natural factors lead to unique climatic zones across New England.
These factors include topography (that is, the mountains, coastal plain,
and other geographic features), ocean currents (both warm and cold),
and variations in daylength and solar radiation. Collectively, these factors lead
to a very complex geographical pattern of climate types across the region.

Topography

Elevation across New England ranges from sea level along the coastal zone to
a maximum elevation of 6,288 feet above mean sea level at Mount Washington
(fig. 1.1). There are several other impressive mountains within the region, including Mount Katahdin, Maine, Mount Greylock, Massachusetts, and Camel's
Hump, Vermont. The major mountain ranges include the White Mountains in
New Hampshire, the Green Mountains in Vermont, the Berkshires in western
Massachusetts and Connecticut, and the Longfellow Mountains in Maine. All
of these mountains are part of the Appalachian Mountain chain of the eastern
United States.

Mountains are notorious for creating their own weather, which is frequently
different from conditions elsewhere within the region. For example, it is not
uncommon to have heavy snow in the White Mountains, while at the same time
conditions are sunny and clear along the seacoasts of Maine and New Hampshire. Likewise, snow events sometimes occur in the Green Mountains while
locations at lower elevations nearby remain precipitation-free or vice-versa.
These differences partly stem from the way that winds are forced around and
over mountains. When a parcel of air is lifted over a mountain, the temperature
of that air cools because pressure is lower and the air is allowed to expand
at higher elevations. When air expands, molecules have fewer collisions and

rebounds with other molecules, resulting in colder temperatures. In turn, the reduction in temperature often leads to condensation of water vapor in the form of clouds and ultimately precipitation. Figure 5.1 shows clouds backed up against Mount Washington.

Perhaps the best example of enhanced precipitation at higher elevation can be found at Mount Washington, which averages 99 inches of liquid-equivalent precipitation (that is, rain plus melted frozen precipitation) in a year and 259 inches of snow. In comparison, Pinkham Notch, located in the valley just below Mount Washington, averages 57 inches of liquid-equivalent precipitation annually and 153 inches of snowfall. This enhanced precipitation is prevalent not only at Mount Washington, but at all the high peaks in the region. There also is an enhancement in wind speeds at higher elevations because of the overall reduction in friction (or drag) from Earth's surface and the compression of the air column as it moves over the mountain. This contributed to the windspeed of 231 miles per hour recorded at Mount Washington, which is perched several hundreds of feet above the other peaks in the Presidential Range. More specific details on these phenomena are given in chapter 12 dealing with the alpine zone.

Another way in which New England mountains affect winds is through the formation of valley and mountain breezes. Under relatively calm conditions,

Fig. 5.1. Example of the orographic effect of the mountains in New England. The view from the Mount Washington Observatory shows undercast clouds backed-up against the mountain. Photo from http://www.mountwashington.org/rotating/february2000/undercast2.html. Used with permission from the Mount Washington Observatory.

Fig. 5.2. Schematic diagram showing how the Labrador (cool) and Gulf Stream (warm) currents meet off the coast of New England (NE). Modified from ATMOSPHERE: AN INTRODUCTION TO METEOROLOGY 7/E by Lutgen/Tarbucks, © 1998 (fig. 3-7, p. 55) Reprinted by permission of Pearson Education, Inc., Upper Saddle River, N.J.

afternoon heating of the mountains tends to produce winds that originate in the valleys and move upslope toward the mountain summits. Frequently, these valley winds produce thunderstorms and can be problematic to day hikers, especially those who venture above treeline in summer. At night, under calm conditions, mountain winds often form by cold air draining from the mountain tops into the valleys. These conditions often form temperature inversions where air at the surface is colder than air aloft, so that the valley floors will be colder than the higher elevations on the valley walls.

Oceans

The climatology of New England is also significantly influenced by ocean proximity. The southern shore of New England, including Connecticut, Rhode Island, and southern Cape Cod, has relatively warm water offshore, which is modified by the Gulf Stream, a warm water current flowing from south to north along the East Coast of the United States (fig. 5.2). In contrast, the eastern shore of New England, including coastal Maine, New Hampshire, and eastern

Massachusetts, is influenced by the cold-water Labrador Current flowing southward from eastern Canada toward New England. These two currents, which meet near the southeastern tip of Cape Cod, play a major role in the spatial variability of climate across New England.

Along the southern shore of New England (for instance, in New London, Connecticut), sea surface temperatures range from 37°F in February to 73°F in August. The coastal zones of New Hampshire and Maine are substantially colder. In Eastport, Maine, for example, sea surface temperatures are near 35°F in February and 52°F in August and September. The warmer water temperatures and the quicker warming of the ocean heading into summer lead to more successful summer beach resorts on the southern shore of New England, close to Charlestown, Rhode Island, and Hyannisport, Massachusetts, for example, because of comfortable water temperatures for swimming.

Along the eastern coast of New England, the cooler water in summer drives a distinct sea-breeze/land-breeze system that leads to mild summer afternoons (see fig. 2.2). The warm air over the land becomes buoyant in the afternoon hours, and is lifted and replaced by the relatively cool air from over the Gulf of Maine. If you dislike hot summers, the coastal zone, inside of Route 1, would make an ideal location for your residence. The cool afternoon temperatures along the coast also mitigate afternoon thundershowers. In fact, the eastern coastal region of New England receives fewer thunderstorms than anywhere else east of the Rocky Mountains (see chapter 9), a major reason why this area is one of few in the United States that receives more precipitation in winter than in summer (Trewartha, 1981). The snow belt east of the Great Lakes is another area that receives more precipitation in winter than in summer.

Ocean temperatures are colder in winter than in summer, but remain relatively warm compared to land temperatures. This temperature differential in turn has a large impact on the snow climatology of New England, through its influence on the rain/snow line. Frequently, temperatures in the coastal zone are modified by the oceans and remain slightly above freezing during some snowstorms. Hence, a higher proportion of winter precipitation in these areas falls as rain or freezing rain as compared to areas farther inland (see chapter 15). This effect is more pronounced along the southern shore of New England where sea surface temperatures are even warmer.

The impact of the oceans can also be thought of as a shock absorber to climate extremes, because ocean temperatures change very slowly. In other words, the coasts are warmer in winter and cooler in summer than inland locations. For example, Burlington, Vermont, averages 16.3°F and 70.5°F in January and July, respectively. At Portland, Maine, located at roughly the same latitude, mean temperatures for the same months are 20.8°F and 68.6°F, respectively.

Daylength and Solar Radiation

Seasonal changes in temperature are also controlled by the length of daylight and by solar intensity. Temperatures are warmer in summer because the number of daylight hours is longer and the sun's rays are more direct than at any other time of the year. At the central New England location of Concord, New Hampshire, the length of daylight ranges from 15 hours and 24 minutes on the summer solstice (June 21–22) to 8 hours and 59 minutes on the winter solstice (December 22–23). The intra-annual change in day length is even greater north of Concord and less so to the south, as given in table 2.1. On the summer solstice at Concord, the sun climbs higher in the sky than at any other time, peaking near 70° above the horizon at solar noon (where 90° would have the sun directly overhead), resulting in high intensity radiation at the surface. On the winter solstice at Concord, in contrast, the sun only rises to near 23° above the horizon.

Summary

The combination of three factors—elevation, proximity to the ocean, and latitude—create most of the geographical patterns of weather and climate in New England. A decrease in temperature and an increase in windiness (both frequency and magnitude) occur with elevation, thus mountainous regions are cooler than the surrounding valleys. Mountains also act as barriers to air flow, and, with the cooling of air as it rises over a mountain, condensation and precipitation occur. Precipitation will be greater on the windward side of the mountain than on the lee side. The influence of the ocean on weather and climate in the region is particulary marked because of the contrasting cool Labrador Current off the eastern coast and the warm Gulf Stream off the southern New England coast. Temperatures are moderated by the cooler current. The cooler air also helps inhibit thunderstorm activity in coastal regions of New England. The maritime influence produces warmer temperatures along the coast during winter. Latitudinal variation in solar radiation and hours of daylight further influence temperature regimes across the region, specifically the cooler temperatures in northern Maine versus the warmer temperatures in southern Connecticut.

Chapter 6

Nonpermanent Causes

Rarely will you hear a native describe a season or year as being typical.
—BENJAMIN WATSON

The factor that causes the greatest variability in New England's day-to-day weather is the positioning of the high (anticyclone) and low (cyclone) pressure systems. When the position of these systems follows consistent patterns over time, these transient features can have a great influence on average conditions for a particular year (that is, the climate). We will describe the types and locations of the major weather patterns that influence New England and the time of year when these systems are most prevalent. Many of these systems have been given specific names that ease their identification, particularly for purpose of discussion. The names often are based on the track the particular system takes or its primary location (such as the center of a high pressure system) or the location of its origin. Identifying a pressure system by where it formed—that is, at the point of cyclogenesis or birth of a cyclone—is especially prevalent in the case of low pressure systems.

We will first describe the major high pressure systems that influence New England's climate and then the major low pressure systems and the tracks these storms often follow. In both cases, we briefly describe the impact on New England's weather and climate. The last part of the chapter discusses how certain circulation systems in the northern hemisphere relate to each other (teleconnections) and how these relationships affect New England.

High Pressure Systems

There are two times of the year when specific high pressure systems exert an exceptionally strong influence on New England's weather. One time is during the winter, when a large dome of high pressure forms over north-central Canada. This cold, dense air mass remains over that area for extended periods of time with smaller high pressure centers moving from it into the northeastern

United States. The persistence of this anticyclone is aided by snow cover in northern Canada, since the cold snow surface maintains both the cold temperature and the stability of the overall air column.

This large dome of high pressure migrates across the northern latitudes during the year, and it is often situated close to Hudson Bay in February, hence the term Hudson Bay High (Ludlum, 1976). This high pressure cell provides the contrast between the colder air over snow cover and warmer air over the oceans or more southern land masses of New England that is the ideal ingredient for the development of the large winter storm—nor'easters—as discussed in chapter 15, although if the high pressure center is very strong it can keep storms south of New England.

Air masses originating from the Canadian high pressure system bring very cold polar air into New England on northwest winds (fig. 2.10). At times, air will be drawn across the North Pole from northern Russia, thereby pushing very frigid air into New England. These individual high pressure systems influencing New England are often referred to as the "Siberian Express." The coldest nights and days occur when these high pressure systems drift directly over New England. The combination of lack of wind at the center of the high preventing the mixing of air, the loss of daytime heating with setting of the sun, and the loss of heat by radiational cooling makes for frigid nighttime temperatures.

During other times of the year, cold air originating in central Canada will migrate into New England. Although not as frigid as winter air masses, flow from central Canada is often quite cool compared to the air mass it replaces. This chilly air also arrives on northwest breezes on the eastern side of the high pressure system (fig. 2.10). However, once the high migrates to the east or southeast of New England, flow shifts to the southwest, resulting in the advection of warmer air into the region (fig. 2.11). On occasion, the high pressure system moving through New England will pass to the northeast, bringing flow off the Atlantic Ocean. This type of flow produces cool, humid conditions, and is often associated with fog along the coast (fig. 2.12).

The second time period when high pressure systems may be dominant is during the summer. At this time of the year and occasionally during other seasons, high pressure systems will "park" themselves to the south or southeast of New England (fig. 2.11). The southwest flow around these highs can produce very long heat spells in New England. Because the flow is coming essentially from the Gulf of Mexico or subtropical Atlantic Ocean, these air masses are also quite humid, with dew point temperatures in the high 60s and possibly into the low 70s. Dew point is the temperature at which water vapor condenses. Dew points above 60°F are responsible for "sticky" conditions, and those in the 70s are considered tropical. In this case, summer forecasts are best summarized by the "3 Hs," that is, hazy, hot, and humid. This is often the best time for going to the beach or to a lake, but not much else. The high pressure system that moves to the east of New England is basically migrating toward and reinforcing the circulation of the semi-permanent Bermuda-Azores High of the subtropical Atlantic Ocean. The Bermuda High in summer is notorious for serving as a "heat pump," driving very hot and humid air from the Gulf of Mexico and subtropical Atlantic Ocean all the way to New England.

Storm Tracks

New England occupies a unique position as far as how storms (low pressure systems or cyclones) migrate across the country. Most extratropical storms reach the western coast of the United States, many forming in the northern Pacific particularly around the Gulf of Alaska. This especially occurs when the Aleutian Low forms in that area of the Pacific Ocean. In fact, most well-developed storm systems form in winter; slightly fewer form in the transition seasons of spring and fall. These are the periods (particularly during the winter) when temperature contrasts between latitudes or between the land and the ocean are at a maximum. A high contrast in temperature between areas that are close together provides fuel for the generation of cyclones—a process called cyclogenesis. Well-developed storms are not as frequent during summer because temperature contrasts across the United States are minimal. Once a well-defined system forms, it will eventually head in the general direction of New England. Storms take this path because these systems are migrating toward the Icelandic Low, the semi-permanent subpolar low pressure system in the northern Atlantic. Thus, almost all storm tracks converge on New England, thereby placing the region in a unique position as the tailpipe for weather systems in the United States (fig. 6.1).

We summarize below the major storm tracks as they appear across the United States. We divide the discussion into two parts. The first describes the storm tracks that approach New England from the west to northwest. The second describes systems that approach New England from the west to southwest.

West-Northwest Approaching Storm Tracks

Storm systems that move across the northern tier of the country toward New England usually are weaker than their counterparts that migrate across the south. Storms that move almost directly east into New England are removed from the major moisture source of the Atlantic Ocean and Gulf of Mexico, thus they often will not produce copious amounts of precipitation. They also do not tap as much humid air from the south, which reduces temperature contrasts and limits the size and strength of the storm. However, these storms can produce moderate amounts of precipitation in the New England region.

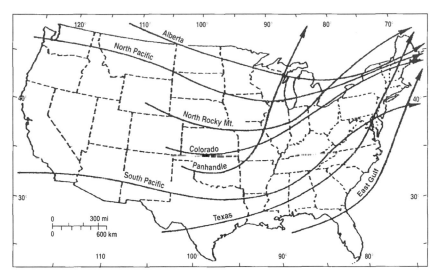

Fig. 6.1. Common storm tracks across the United States showing how most tracks converge upon New England. From the New England Regional Assessment Group (2001; fig. 1.1, p. 3), University of New Hampshire, after METEOROLOGY: 5/E, THE ATMOSPHERE AND THE SCIENCE OF WEATHER by Moran/Morgan © 1997 (fig. 11.16, p. 274). Reprinted by permission of Pearson Education, Inc., Upper Saddle River, N.J.

North Pacific. These storms originate off the coast of Washington state, and they are able to stay intact as they migrate across the Rocky Mountains. They generally will track through the upper Midwest and may pass through northern New England. If a distinct trailing cold front lies south of the low pressure system, northern New England may receive precipitation followed by cooler temperatures. The key to storms that follow this track is that they must stay intact as they cross the northern Rocky Mountains.

Alberta Clipper. Although these storms "originate" in Alberta, they actually form in the northern Pacific. As the storm crosses the much higher landmass of the Rocky Mountains, it is compressed vertically, while at the same time it expands horizontally. The result is that the original storm loses much of its form and identity. Once the original storm moves onto the Great Plains to the east of the Rocky Mountains, the storm reforms or re-energizes. Such storms are referred to as Alberta Clippers when this process occurs in the southern part of Canada (fig. 6.1). This is the same process that occurs in the development of the Colorado Low, to be discussed below. Alberta Clippers often are fast-moving systems that dip initially into the northern United States and move across the Great Lakes. They eventually migrate into New England and are more frequent in winter than in other seasons. They are usually intense enough to produce about 4 to 6 inches of snow or possibly up to about 8 inches, but distribution of that snow is often limited. Greatest snowfall amounts will be close to the storm's center, with snowfall totals tapering off dramatically away from the main path of the storm. Sometimes these storms will intensify off the New England coast, but areas of heavy precipitation frequently are well off the coast or they may catch Down East Maine and the Canadian Maritimes. The manner in which the Blizzard of '78 formed is an exception to this scenario (see chapter 15).

West-Southwest Approaching Storm Tracks

Storms originating more toward the southern boundary of the United States are often more intense than storms that form close to the U.S.-Canadian border, because of the great temperature contrast between the land and the Gulf of Mexico and Atlantic Ocean waters off the southeast coast. This contrast is enhanced during winter. As a result, many storm tracks move across the southern part of the United States and up the eastern seaboard, eventually entering or passing near New England.

South Pacific. Much like the Northern Pacific storm track, the southern track consists of systems that are able to maintain their form as they cross the southernmost Rocky Mountains (fig. 6.1). Eventually these storms may move across the southern states before turning northward along the Atlantic Coast. Sometimes, warm and humid air from the Gulf of Mexico gets entrained into these systems.

Colorado Lows. In many ways, the Colorado storm track is the southern equivalent of the Alberta Clipper, with a major difference. The area where Colorado Lows form is very active in winter and it is an area of strong cyclogenesis. This is such a prevalent area of strong storm formation because storms are able to tap into the moisture and energy available from the Gulf of Mexico once they re-form after moving across the Rocky Mountains. These storms can become quite strong while producing copious amounts of precipitation once they reach New England, especially when they tap Atlantic Ocean water with migration toward the Northeast. However, as storms forming in Colorado move to the northeast, they may take two main tracks. One track is northward across western New England, eventually moving down the Saint Lawrence River valley. The second possible track is along the New England coastline or slightly offshore. As every New Englander knows, when it comes to winter snow storms, the track is all-important in determining whether a particular area will receive snow exclusively, snow turning to rain, mixed precipitation, or all rain. We elaborate on this aspect of nor'easters in chapter 15.

Gulf Coast Lows. The Gulf Coast low pressure system often forms in the western part of the Gulf of Mexico (fig. 6.1). The low pressure systems originating here feed off of the contrast between the land mass of the mid-continent and the warmer waters of the Gulf of Mexico. These storms can be especially strong when snow cover extends very far into the Gulf Coast states, thereby providing that extra contrast in temperature. In fact, when snow cover does extend into the southernmost states, the winter-time Arctic front will follow it. The Arctic front is the equivalent of the polar front discussed in chapter 2, although the Arctic front in North America is primarily a wintertime phenomena (fig. 2.6). It marks the contrast in temperatures between the cold air above the snow-covered regions of the North American continent and the snow-free areas south of them. This was the scenario that helped generate the March 1993 Superstorm. As was the case for the Colorado Low, storms forming in the Gulf of Mexico can become quite intense. They can also take one of the two major tracks upon moving toward the Northeast (i.e., up the west side of the Appalachians eventually moving through the Saint Lawrence River valley or up the east side of the Appalachians along the Atlantic Coast). Whether they originate as a Colorado or a Gulf Coast Low, storms moving up the Atlantic Coast can undertake another round of cyclogenesis or strengthening. When this occurs, our final storm track comes alive.

Atlantic Coast. A very important factor that contributes to the re-formation of storms or to the rapid intensification of storms off the Atlantic Coast is the Gulf Stream (fig. 5.2). This warm ocean current and the air above it, again, provides the tremendous contrast to the colder air over the continent. Thus, many storms reaching the Atlantic Coast from October to April have the potential to form into major nor'easters.

Teleconnections

The word "teleconnections" conjures up thoughts of calling a relative or friend on the phone or via email. Interestingly, that is essentially what the word describes from a climatological perspective. Simply put, a teleconnection describes how various weather conditions that can be half the world away from each other are related. Very often the nature or intensity of a teleconnection pattern is stated by a single number or index to make it easier to identify. This single number may be calculated by subtracting one pressure from the other or from a series of pressure readings. Whether the result is negative or positive is indicative of certain conditions within the region defined by the particular teleconnection. We now give specific examples of teleconnections that play a major role in year-to-year variability in the New England climate.

The North Atlantic Oscillation. The teleconnection that plays perhaps the most important role in determining the weather and climate of New England involves the strength of the Icelandic Low relative to that of the Bermuda-Azores High. This relationship is referred to as the North Atlantic Oscillation (NAO) because the actual value for the oscillation is calculated by looking at the pressure difference between Akureyri, Iceland, and Punta Delgada, Portugal, in the North Atlantic. The relationship between the pressures at these two sites produces either a negative phase of the NAO or a positive phase of the NAO (fig. 6.2). When both the Icelandic Low and the Bermuda-Azores High are very strong, there is a strong pressure gradient between the two. This positive mode of the NAO produces a jet stream that flows strongly between the two circulation systems. This produces a zonal flow in the upper-level westerlies across New England. When both the Icelandic Low and Bermuda-Azores High are weak (that is, a higher pressure for the low and lower pressure for the high), a high pressure system tends to form over Greenland, producing a blocking pattern over the North Atlantic (see fig. 2.9). This scenario produces a meridional flow pattern across the eastern United States and New England. When this occurs, a strong contrast of air masses across the northeast provides the fuel for storm development. Remember, meridional flow indicates greater temperature contrasts and more storm activity, while zonal flow produces smaller temperature contrasts and less storm activity.

The NAO exerts most of its influence on New England during the winter. When the NAO is in the negative mode, the more frequent meridional flow will produce a greater number of coastal storms and the influx of colder air into the eastern part of the country: a prime mix for heavy and frequent snow storms. The actual number of storms and snowfall totals will depend on exactly where the limb of the jet stream is relative to New England. If it is frequently well to the east of the region, New England will have a cold winter, but most storms will move off shore. If it is essentially over New England (that is, along the coastline to the western edge of New England), the major storm track should be close enough to New England that we could have a big snow year. If it moves too far west, the region would be on the warm side of the storm track and if the area is impacted by storms moving down the Saint Lawrence River valley, they most likely will be rain events. If the NAO is in the positive mode, the zonal flow of the jet stream will bring modified Pacific air into New England, causing a cool to warm winter without major coastal storm development. However, it is important to realize that the NAO will vary within a year or season, so that we can experience a major snowstorm or two even in winters when the NAO is in a positive mode for most of the season.

Pacific–North America Pattern. This teleconnection pattern is rather interesting; the way it is calculated produces a number that determines whether the coterminous United States is dominated by meridional or zonal flow. The Pacific–North American (PNA) reflects circulation patterns upwind of New

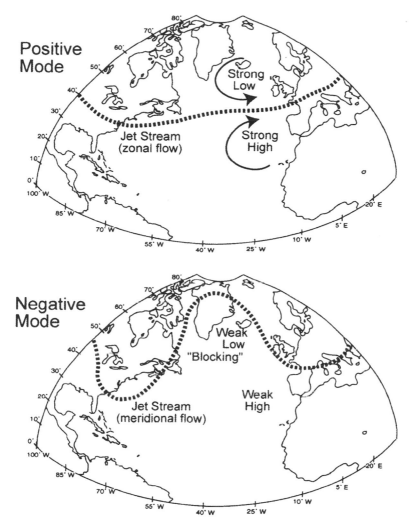

Fig. 6.2. The two modes of the North Atlantic Oscillation (NAO) and their influence on the shape taken by the jet stream. As explained in the text, the meridional flow associated with the negative mode of the NAO (*bottom*) often leads to a greater number of winter storms and snowfall in New England. From the New England Regional Assessment Group 2001. *Preparing for a Changing Climate: The Potential Consequences of Climate Variability and Change.* New England Regional Overview. U.S. Global Change Research Program, 96 pp., University of New Hampshire.

England, whereas the NAO reflects circulation patterns on the downstream side. The PNA also reflects the strength of the Aleutian Low. In a simple sense, the PNA is determined by comparing the pressures just off the coast of Alaska, in Alberta, Canada, and in the southeastern United States. When the differences among these individual sites is negative, zonal flow prevails across the country, while there is meridional flow when the PNA is positive (Leathers et al., 1991). Conditions in New England will vary accordingly, as we have described previously.

Other teleconnections in the northern hemisphere affect New England more indirectly than these two. These are summarized on the Climate Prediction Center, NOAA home page for northern hemisphere teleconnections at www.cpc.ncep.noaa.gov/data/teledoc/telecontents.html, at the time this was written.

Summary

Two primary high pressure systems play dominant roles in how New England's weather may vary over time. Large high pressure systems over central northern Canada feed cold air into New England on brisk northwest winds. High pressure systems moving off this dense dome of cold air during the winter are responsible for the very cold winter temperatures in the region, including long cold spells. The presence of snow cover in these areas of central Canada throughout the winter adds to the stability and cold surface temperatures associated with these anticyclones. Air masses originate from these areas in other seasons, causing cooler temperatures. During summer, high pressure sometimes becomes fixed south to southeast of New England as part of the Bermuda-Azores High. These systems pump in hot, humid air and are responsible for those excruciating hot, humid, and hazy days of summer—the dog days—and to lengthy heat waves. Flow of this nature occurs in other seasons, producing warm or mild temperatures.

Low pressure systems migrate across the United States along several prevalent storm tracks, most of which converge on New England. Many of these systems originate to the east of the Rocky Mountains as storms re-form after being distorted by the high elevations of the Rockies. Alberta Clippers move across the northern United States bringing only moderate amounts of precipitation to New England. Areas receiving the higher amounts of precipitation are usually those close to the path of the storm. Colorado Lows and Gulf Coast Lows, on the other hand, receive energy input from the warm waters of the Gulf of Mexico and have the potential to intensify into very large, strong storms. Great temperature contrasts, which primarily occur between ocean waters and land masses during winter, provide fuel for the generation of large storm systems along these tracks. Storms forming in Colorado or in the Gulf of Mexico can move northeastward along the western side of the Appalachians and down the Saint Lawrence River valley or along the eastern side of the Appalachians, closer to the Atlantic coastline. The track determines which parts of New England will be on the warm, east side of the central low pressure or on the cold, west side of it. Storms tracking up the Atlantic Coast may intensify or re-form with the aid of the warm waters of the Gulf Stream, thus having the potential to turn into monster nor'easters.

The interactions among different circulation systems, referred to as teleconnections, also have an influence on New England climate. The North Atlantic Oscillation (NAO) indicates the relative strength of the Icelandic Low compared to the Bermuda-Azores High. It probably exerts more influence on New England weather and climate than any other teleconnection, especially during the winter. In the positive mode, that is, when both systems are strong, a strong zonal flow forms across the United States and New England. This results in cool to mild winters and a lack of significant snow storms. However, when the NAO is in the negative mode because both systems are weak, a high pressure cell will form over Greenland, producing a blocking pattern and a strong meridional flow pattern across the eastern United States. This pattern is more likely to bring colder and snowier winters to New England, as it is ideal for the development of nor'easters. The Pacific–North American (PNA) teleconnection is indicative of zonal or meridional flow across the contintental United States, with similar ramifications to that of the NAO.

Part IV

Seasons of New England

In the spring I have counted 136 different kinds of weather
inside of four and twenty hours.   —MARK TWAIN

In the preceeding three parts of this book, we introduced the many factors that control New England's weather and climate. Now we will take you through a "typical" year in New England (if such a year existed!), presenting the conditions produced by the combination of all of these factors. We describe the general temperature and precipitation patterns that occur in each season, given the individual controls that exert the most influence during specific times of the year.

However, the seasons that we present are those defined by New Englanders themselves, not the conventional winter, spring, summer, and fall. Moreover, New Englanders have defined sub-seasons within the major seasons that also are highly influenced by weather and climatic conditions. New Englanders know that particular aspects of the seasons so greatly influence everyday life that the typical four-season sequence may not be the most appropriate way to define them.

Two recent examples epitomize how we believe New Englanders characterize the seasons. Although both of these examples are related to spring, the general idea holds for all seasons. The first example comes from the "Dear Yankee" section in the July 1999 issue of Yankee Magazine. Jeanne Celeste of Hudson, Massachusetts, wrote a letter to comment on driving in Vermont near the end of winter or early spring. Upon seeking help at a farmhouse after getting her car stuck in the mud, she was promptly greeted by, "You shouldn't be out during mud!" Clearly, the woman who uttered those words knows only too well how to define the seasons of New England.

Another example of how New Englanders define the seasons arose during the planning for New Hampshire's exhibit at the 1999 Smithsonian Folklife Festival. One particular meeting raised the question, "how do you characterize spring in New Hampshire?" The answer was short and to the point: "mud and blackflies." In other words, once the snow starts to melt, it is mud season; within mud season comes black fly season.

Of course, the great variability in weather and climatic controls across the region mean that the timing of the seasons will vary from northernmost Maine to southwestern Connecticut, as well as from the Northeast Kingdom of Vermont to the outer part of Cape Cod. Consequently, when we identify the months that comprise these seasons, we are making broad generalities regarding their timing.

As such, we define the seasons of New England in the following manner. We will discuss Ski Season (chapter 7), followed by Mud Season (chapter 8).

Beach and Lake Season (chapter 9) will follow, and we end with Foliage Season (chapter 10). The final two chapters of this part of the book present the Year in Summary (chapter 11) and a discussion of the special aspects of the Alpine Zone (chapter 12). Within each chapter, we define the timing of the particular season and the major conditions within each. This part of the book should give visitors or new arrivals to New England a better understanding of the true seasons of New England.

Chapter 7

Ski Season

A heavy November snow will last until April.
—NEW ENGLAND PROVERB

As temperatures tumble and snow begins to fill the air, one's thoughts certainly can turn to skiing, as well as snowmobiling. Ski season can coincide with astronomical winter, but as New Englanders know, winter is the longest season of the year. Depending on where you are in the region, winter can extend from early November to late April—an amazing six months of the year. On average, it may be more accurate to say that the ski season extends from late November to early April. Nevertheless, that still leaves at least five months of the New England year under winter or winter-like conditions. The entire five to six months may not be characterized by winter conditions, but below freezing temperatures and snow can occur just about any time from November through April. Such conditions can even occur in October or May. Several times over the last two hundred years, snow fell in June as far south as the Vermont-Massachusetts line and killing frosts occurred in both July and August.

To describe the main characteristics of New England's winter climate, we will first present the variability in temperature across the region followed by a discussion about the variability in winter precipitation. This will be followed by a section detailing sub-seasons within the ski season.

Temperature

There are two main reasons for the very low winter temperatures found in New England. One is the lower amount of solar radiation reaching the region compared to other times of the year given the very low angle of the sun (fig. 2.3). The second reason is the smaller amounts of daylight overall, particularly during the earlier to middle parts of the season (fig. 2.4; table 2.1). The low temperatures that arise from these factors can be augmented by the prevalence of

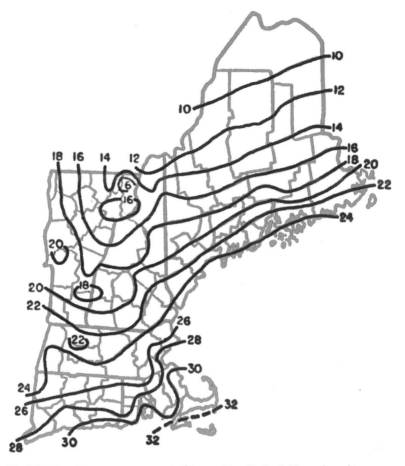

Fig. 7.1. Mean January temperatures in °F across New England. Lines of equal temperature are in 2°F increments. Courtesy of Bob Adams.

large Canadian high pressure systems that transport very cold air from northern Canada to New England. Add snow cover to the mix, and we have the makings for a very cold ski season, on average.

Temperatures warm dramatically going from northern New England to the southern coast of Connecticut and Rhode Island regardless of whether one looks at mean temperatures for the months during the ski season or average maximum or minimum temperatures. Average monthly temperatures vary from about 6°F for January in northernmost New England to 30°F in the south (fig. 7.1). This great difference is a function of the latitudinal distance from northern New England sites such as Caribou, Maine, Saint Johnsbury, Vermont, Berlin, New Hampshire, to places along the coasts of Connecticut and

Rhode Island. Insolation is slightly greater in the south, and at the same time, the ocean modifies (that is, warms) temperatures along these coastlines. Average maximum temperatures in January range from near 20°F in extreme northern New England to 38°F in southern New England. Average minimum temperatures in January range from −8°F to 22°F from north to south, but the timing in the occurrence of minimum temperatures differs across the region. Most inland regions will reach their lowest temperatures around mid-January, whereas coastal sites tend to reach their minimum values around the last week in January or the first week in February (fig. 7.2).

Average daily temperatures at the beginning of the ski season (early November) vary from 36°F in Caribou, Maine, to 46°F in Hartford, Connecticut. Temperatures across New England at the end of ski season (late April) range from 45°F in Caribou to 54°F in Harford. Overall, temperature differences during these transition periods into and out of the ski season change by about 10°F from north to south across the region.

Spatial variability across the region is further highlighted by extreme temperature recordings, such as record low temperatures (table 7.1). Interestingly, these record lows exceed those recorded in Anchorage, Alaska, further emphasizing the variability in New England's climate from the perspective of extreme conditions nationwide. Moreover, the area in the country with the greatest day-to-day temperature variability in the forty-eight contiguous states spans northwestern Vermont and northeastern New York—around Lake Champlain (Calef, 1950). Daily minimum temperatures may vary by an average of 10°F from one day to another in January. These large day-to-day shifts are a function of the migration of a large high pressure system from west of New England (cold northwest air flow) to east of it (warm southwest air flow), followed by the next Arctic blast.

One way to evaluate winter's impact on everyday life is to determine the

Table 7.1
Record low temperatures officially recorded
for each state in New England

State	Location	Date	Temperature (°F)
Vermont	Bloomfield	30 Dec. 1933	−50
Maine	Van Buren	19 Jan. 1925	−48
New Hampshire	Pittsburg	28 Jan. 1925	−47
Massachusetts	Chester	22 Jan. 1984	−40
Connecticut	Falls Village	16 Feb. 1943	−32
Rhode Island	Kingston	11 Jan. 1942	−23

Adapted from Ludlum (1976) and Watson (1990).

Daily Maximum Temperature

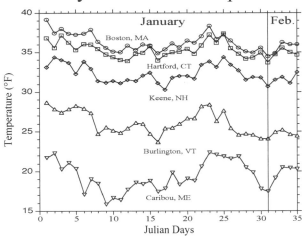

Daily Minimum Temperature

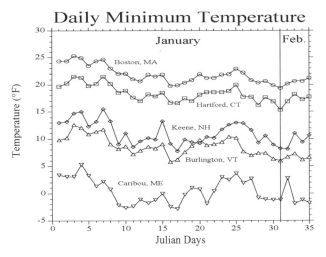

Fig. 7.2. The top graph shows mean daily maximum temperature for January and the first four days of February for several cities across New England. Note the peak in maximum temperature between 22 and 25 January for each city compared to the surrounding weeks. The bottom shows mean daily minimum temperature for January and the first four days of February for several cities across New England. The peak in minimum temperature between 22 and 25 January is not as distinct as it is for maximum temperature. Similarly, a plot of mean daily temperatures would not show the January thaw around 22 to 25 January as distinctly as does mean maximum temperature. Also note that lowest temperatures for inland sites such as Caribou, Burlington, and Keene occur around 17 January, whereas coldest temperatures for southern and coastal sites such as Hartford and Boston occur around 31 January. Mean values were calculated for the period 1920 to 1999 for Boston, Hartford, and Burlington, 1926 to 1999 for Keene, and 1941 to 1999 for Caribou.

number of heating-degree days (HDD) occurring during the winter season. This index was developed to reflect the coldness of the winter as it relates to energy expeditures to heat one's home. Heating-degree days are calculated by cumulatively counting the number of degrees the mean daily temperature is below 65°F throughout the winter, thereby producing a number that reflects how much one has to spend on heating fuel. An average day during the heart of winter will produce about 60 heating-degree days in northern New England, and 35 heating-degree days in southern New England. An average winter accumulates over 9,000 HDD in northern New England and under 6,000 HDD in southern New England (fig. 7.3).

An interesting phenomena during ski season is the presence of a climatic singularity in New England. A climatic singularity is defined as a particular climatic event that occurs every year at roughly the same time. In New England, and perhaps other areas of the Northeast, the most widely known cli-

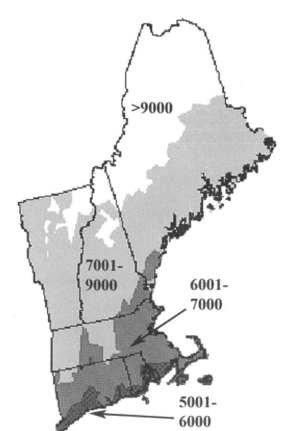

Fig. 7.3. Median number of heating degree days (HDD) across New England. Numbers represent the number of HDD during the heating season that one would expect 50 percent of the time. Calculation of HDD is described in the text. Modified from the *Climate Atlas of the United States* (National Oceanic and Atmospheric Administration/Department of Commerce, 2000; disk 1).

matic singularity is the January thaw. Almost every January, often between the twenty-first and twenty-fifth day of the month, mean temperatures will be higher than those during the weeks immediately preceding or following these days. Also, maximum daily temperatures around 22 to 25 January over much of New England are noticeably higher than the maximum temperatures on the days before and after this period (fig. 7.2). Daily minimum temperatures do not show this distinct warming over these same days. The actual maximum temperature does not necessarily fall on the same day in January throughout the region, but it is almost always during the third week in January everywhere. Also, it should be noted that the term "thaw" can be misleading, because areas such as Burlington, Vermont, and Caribou, Maine, will often have maximum daily temperatures below freezing during those days, despite the fact that those days are frequently warmer than during the week before and after.

The January thaw appears to be caused by a northward and westward shift in the Bermuda-Azores high pressure system of the subtropical Atlantic Ocean (see chapter 2 for a discussion of the relevance of this high pressure system to New England's climate). This shift in the position of the anticyclone allows a more southerly to southwesterly flow to reach New England, thereby producing days with higher temperatures than the adjacent days—the January thaw. Climatologists are uncertain why the Bermuda High migrates in this manner during the third to fourth week of January. However, New Englanders are also aware of the fact that not every January has a true thaw, as evidenced by the deep snowpack that persisted through the winter of 2000/2001.

Precipitation

Winter precipitation in New England includes rain, snow, and every combination in between. Most winter precipitation is generated by mid-latitude cyclones, including Alberta Clippers, Colorado Lows, Gulf Lows, and Atlantic Lows (fig. 6.1). When these storms are situated close to the coastline, large amounts of precipitation may be generated by flow off the Atlantic Ocean on northeast winds circulating counterclockwise around the central low pressure, hence the term nor'easter is used to identify such storms (fig. 2.14; see also chapter 15). However, the track of the storm will determine whether a particular site will receive rain, freezing rain, sleet, snow, or a combination of any to all of these types of precipitation. This scenario is what makes forecasting winter storms in New England so challenging.

Several key factors influence the nature of winter precipitation in New England. As we are especially interested in the ski season in this chapter, we will focus on the solid component of winter precipitation: snowfall amounts. Long-term snowfall averages are greatly influenced by two specific factors, latitude

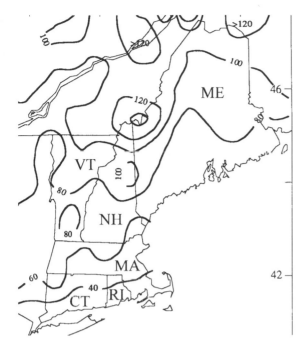

Fig. 7.4. Median seasonal snowfall across New England, in inches. Snowfall totals represent the total amount of snow per season one would expect to receive 50 percent of the time. This number is likely to be lower than average snowfall totals because extreme or very high snowfall events will often increase the average amount for a particular site. Contours are 20 inches. Period of record used to develop this map is 1955 to 1992. Modified from Cember and Wilks (1993; map 49) as available from the Northeast Regional Climate Center (National Oceanic and Atmospheric Aadministration/Department of Commerce), Cornell University.

and elevation. On average, northern New England is colder compared to southern parts of the region thus more precipitation will fall as snow (fig. 7.4). Interior New England is mostly mountainous, thus these high elevation locations are cooler. Also, topography enhances snowfall accumulation as air is forced to rise up a mountain. As air rises, it cools. Precipitation often results since colder air cannot hold as much moisture as warmer air. Areas on the windward side of the mountain range will receive greater amounts of precipitation than sites on the leeward side.

On a year-to-year basis, precipitation and especially snowfall amounts within the region vary quite markedly (fig. 7.5). The main cause is often the nature of the prevalent storm tracks throughout winter. Three primary scenarios come into play. In one scenario, New England is under the consistent influence of the large anticyclone in northern Canada. This will keep cold air in place over the region, but it is a very stable air mass. The jet stream, the zone of very fast upper-level winds, thus will be located south of New England. This produces a cold winter, but not necessarily a snowy winter, since many storms will be forced to the south of New England. From a recreational point of view, this may not be the worst scenario, because downhill ski areas will be able to make snow given the cold conditions in place. Similarly, lakes and ponds will develop a very thick ice cover without snow to insulate the

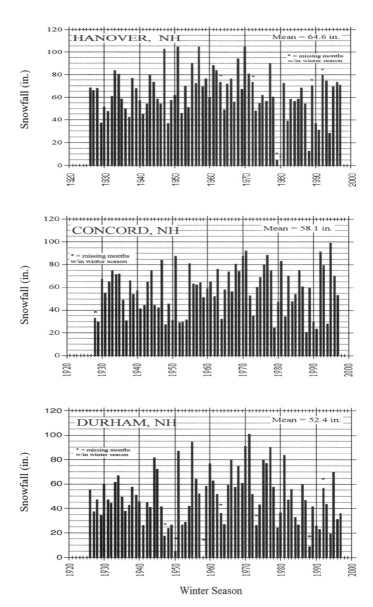

Fig. 7.5. Amount of snowfall received per winter season for the 1920s to 1997 for Hanover (*top*), Concord (*middle*), and Durham (*bottom*), New Hampshire. The variability in amounts from the far western site of Hanover to the central Concord site to the near coastal Durham site is representative of how snowfall amounts vary across these same regions throughout all of New England. Note the amount of year-to-year variability in snowfall amounts, particularly in the record from the near-coastal site of Durham.

water from cold air temperatures. This provides excellent conditions for ice skating and ice fishing.

The second scenario that influences type and amount of precipitation is the presence of a very strong zonal flow (west to east) across the United States. Several possibilities can occur given this situation (fig. 2.9). With zonal flow, and the jet stream north of New England, storm systems may frequent northern New England. This situation could be characterized by Pacific air and mild conditions. If the jet stream is south of New England, the region will be under the influence of cool and dry air, but not as cold and dry as conditions when the Canadian high is in place. If the jet stream is over New England, northern sections will be cold, with mild conditions to the south. Alberta Clippers often are associated with this last scenario, producing snow in the north and rain in the south.

The third overall winter scenario is a strong meridional flow in the form of ridges and troughs across the country (fig. 2.9). Troughs in this circulation consist of lobes of cold Canadian air displaced into the United Sates. Ridges are made up of warm tropical air pushed up into mid-latitude locations. Areas beneath ridges are warmer than normal and areas beneath troughs are cold. Therefore, regional temperatures are a function of the position of the ridges and troughs. Should the east side of the trough be close to New England, it is likely that storms will migrate along this limb and up the eastern coast. Consequently, exact location of ridges and troughs in the jet stream is key to the type and location of New England precipitation. In one case, the storm track moves up the west side of the Appalachians, as discussed in chapter 6. This pattern will move many storms up through central New England, possibly producing snow in western and northwestern parts of New England, but rain along the coast. In some cases, all of New England may receive rain when the major storm tracks up the western side of the Appalachian Mountains. When the storm track is off the coast, eastern New England may receive heavy snow while lesser amounts of snow or none at all may fall in western New England. Thus, proximity to the ocean is a double-edged sword. It can lead to more precipitation falling as rain in a particular storm or it can mean heavy snow. Overall size and strength of the storm system also is important, as we describe in more detail in chapter 15.

Snowfall amounts vary from around 140 inches in northernmost sections, such as northwest Maine and northern New Hampshire, to less than 40 inches per winter along the southern coast (fig. 7.4). For the period from 1970 to 1999, Bridgeport, Connecticut, averaged only about 25 inches of snow per year. Mountainous regions receive more snow than other areas of the region, with Mount Washington receiving an average of 254 inches of snow per year. At the same time, the number of snow events per winter season will increase moving toward the north, probably due to overall colder temperatures. We define snow

Table 7.2

Average number of snowfall events per year of different magnitudes across New Hampshire from the 1920s to 1998 and a comparison to the historical record

Station					Snowfall amount					
	≥1 in.	≥4 in.	≥10 in.	≥30 in.	<4 in.	4–8 in.	8–12 in.	12–16 in.	16–20 in.	20 in.
INSTRUMENTAL RECORD										
Hanover (1927–1998)	15.0	5.9	1.1	0.0	9.4	3.6	1.4	0.5	0.1	0.1
Average Snowfall: 73.1 in/yr										
Concord (1930–1998)	13.4	5.4	1.0	0.0	7.9	3.7	1.1	0.5	0.1	0.1
Average Snowfall: 63.8 in/yr										
Durham (1927–1998)	11.8	5.3	0.9	0.0	6.8	3.6	1.1	0.4	0.1	0.1
Average Snowfall: 57.3 in/yr										
HISTORICAL RECORD										
Stratham (1738–1801)	13.8	5.0	1.3	0.0	6.1	3.0	0.8	1.0		0.1
Average Snowfall: 51.5 in/yr for entire record. 58.5 in/yr for period 1756–1800.										

events as snow recorded over 1 inch on successive days either via a single storm that extends over multiple days or a succession of fast-moving storms that quickly follow each other and produce measurable snow over several days in a row. On average, southern New England may see about seven to ten snow events per season. This increases to around fifteen per season in central New England, and to about twenty per season, on average, in northern New England (table 7.2). Some years may have more than this average, while other years may have fewer. Areas in southern New England were hit by up to twenty-five events in the record-breaking winter of 1995/1996, or about three times the average number in that part of the region. (See the discussion in chapter 19 on how the historical record is used to suggest changes in New England's climate over the last few centuries.)

To indicate how variable snowfall can be from site to site, we compiled a list of the greatest fifteen snow storms for all stations across New England whose records extend back over the last seventy or more years, that is, from 1999 back into the 1920s. Although the length of record is not the same for each climatic station in New England, it is quite interesting, nevertheless, that only a few storms were recorded at a majority of the sites. The storms shown on table 7.3 were not necessarily the greatest snowfall producers, but they were the most extensive systems geographically. That is, the tracks of these storms allowed snow to accumulate over much of New England. Some of them, however, were very large snow producers. We discuss some of the larger snow-producing storms in chapter 15.

New Englanders are greatly interested in how much snow is actually on the

Table 7.3
Widespread large snowstorms across New England, 1920s to 1998

17–18 February 1952
14–17 March 1956
24–27 February 1969
25–29 December 1969
6–8 February 1978
6–8 April 1982
13–15 March 1984
13–15 March 1993

These storms appeared in records of the top fifteen snowstorms for more individual stations than any other storms during the seventy-plus years from the 1920s to 1998. This does not mean that these storms had the highest snowfall totals, but that they were extensive enough with a storm track conducive to the deposition of all snow across most of New England.

Fig. 7.6. Maximum seasonal snow depth across New England, in inches. Snow depth values represent the maximum depth of the snowpack during the season one would expect to observe 50 percent of the time. This number is likely to be lower than average snow depths because years with very high snowpacks will often increase the average amount for a particular site. Contours are 4 inches. The period of record used to develop this map is 1955 to 1992. Modified from Cember and Wilks (1993; map 56) as available from the Northeast Regional Climate Center (National Oceanic and Atmospheric Administration/Department of Commerce), Cornell University.

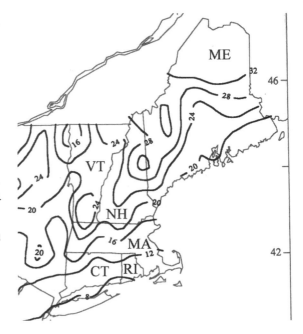

ground at any one time. As was the case for snowfall amounts, the snowpack increases from south to north and from the coast to the interior. Maximum snow depth reached during the ski season varies from less than 1 foot in southern New England to almost 3 feet in the hills of New Hampshire and western Maine and in northernmost Maine (fig. 7.6).

An important aspect of snow depth in New England is the possibility of a "white Christmas," defined as having at least 1 inch of snow on the ground on Christmas day. As expected, the probability of having a white Christmas increases to the north and west. Table 7.4 provides the probability of a white Christmas for several cities in New England. We also were interested in the probability of a "white Thanksgiving," an especially useful piece of knowledge needed for travel by sleigh—as in "over the river and through the woods to grandmother's house we go . . ." (fig. 7.7). Although a white Christmas requires at least 1 inch of snow on the ground, we believe that you would need at least 2 inches to support travel by sleigh (and maps were available for 2-inch snow depths)! We found that most coastal regions will have a 10 percent chance or less of a white Thanksgiving, while northernmost New England may have a 30 to 50 percent chance of having at least 2 inches of snow on the ground, especially when Thanksgiving is in the latter part of the fourth week in November.

Table 7.4
Probability of a white Christmas
in New England

City/Station	Snow Depth		
	1 inch	*5 inches*	*10 inches*
MAINE			
Augusta	90%	52%	21%
Brunswick	80%	40%	17%
Caribou	97%	77%	57%
Houlton	96%	74%	52%
Portland	83%	43%	13%
NEW HAMPSHIRE			
Concord	87%	57%	7%
Lebanon	85%	70%	30%
Mount Washington	93%	63%	27%
VERMONT			
Burlington	77%	50%	13%
Montpelier	93%	50%	13%
MASSACHUSETTS			
Boston	23%	17%	3%
Milton	55%	21%	7%
South Weymouth	33%	13%	10%
Worcester	60%	27%	10%
RHODE ISLAND			
Providence	37%	10%	0%
CONNECTICUT			
Bridgeport	30%	10%	0%
Hartford	57%	23%	3%

Adapted from Ross et al., National Climatic Data Center
technical report 95-03, 1995.

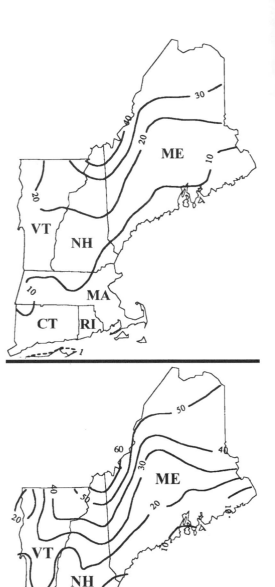

Fig. 7.7. Percent of time when 2 inches or more of snow will be on the ground during 16 to 23 November (*top*) and 24 to 30 November (*bottom*) as an indication of the possibilities for a "white Thanksgiving" (22 to 28 November). For example, if Thanksgiving fell on 22 or 23 November, places such as Boston, Massachusetts, and Portland, Maine, would have less than a 10 percent chance of having at least 2 inches of snow on the ground. If Thanksgiving fell on 24 to 28 November, then Portland would have about a 15 percent chance of having at least 2 inches of snow on the ground. The period of record used to develop this map is 1955 to 1992. Modified from Cember and Wilks (1993; maps 127 and 130) as available from the Northeast Regional Climate Center (National Oceanic and Atmospheric Administration/Department of Commerce), Cornell University.

Fig. 7.8. Sign along a road in Durham, New Hampshire, warning individuals of the presence of frost heaves (*top*). This sign was located just past the sign on the bottom that prohibits the use of vehicles with heavy loads during the period of greatest freeze-thaw and pothole formation. Photos by Greg Zielinski.

Subseasons

Within the ski season, New Englanders identify two overlapping subseasons, sugaring season and pothole season.

Sugaring Season. As the days start to lengthen and warm by late February and early March, the sap in maple trees begins to run. This is the beginning of the sugaring season, just one of the many sub-seasons in New England that is directly influenced by climatic conditions. Great variability in temperature conditions during the sugaring season can lead to periods of either enhanced or more subdued flow. Ideal conditions are reached when the temperatures drop to the mid-20s at night, but then reach highs in the 30s during the day (according to veteran sugarers Bob Moulton and Roy Hutchinson). However, if temperatures get too warm, too quickly, or it is too windy, too quickly, then the sap will stop running. Similarly, if it gets too cold late in ski season, the sap will stop flowing. Interestingly, the key to a good sap season occurs during the previous beach and lake season. Adequate precipitation and sunshine between June and September mean a good crop of leaves from where the sugar is made.

Pothole season. At the same time as maple sap starts to flow, other parts of the landscape begin to thaw. One area of particular interest is beneath roadways. The end of the ski season also increases the frequency of freezing and thawing in the upper few meters of the soil. Cold, dry winters produce a greater thickness of frozen ground that needs to thaw, whereas mild or very snowy winters will produce a thinner portion of frozen ground. When ice forms, the ground expands or heaves (fig. 7.8). With thawing, the ground returns to near original pre-season conditions, leaving voids in the substrate where ice used to be during the winter. Surface material, including the roadway, collapses into these voids, producing potholes, particularly with heavy vehicle traffic. The freeze-thaw process is also responsible for the "appearance" of new rocks in fields and yards each spring. Finer-grained soil returns more rapidly to previous levels upon thawing than do larger particles. The smaller particles end up beneath the larger particles, thereby preventing them from returning to their former levels with contraction during the thaw process. The result is that the larger particles are "pushed up" more and more each winter, eventually reaching the surface. This process is the secret to bumper crops of "New England potatoes" through the ages.

Note that both the sugaring and pothole subseasons may extend into the mud season in any individual calendar year.

Chapter 8

Mud Season

Mud is weather as much as snow.
—DONALD HALL

M ud season, the bane of travel in many rural areas of New England, is a yearly phenomenon that truly indicates that spring is fast approaching. Timing of mud season varies from year to year depending on the amount of winter snowfall and on winter and spring temperatures. Greater snowfall means a greater supply of water to infiltrate the ground upon spring melting, and with it, a more lengthy and severe mud season. Colder winters without much snowpack lead to greater ice formation in the thin glacial soils of the region and abundant availability of moisture for "making mud," although a deep snowpack may provide a more intense mud season. The colder the spring, the later into April and May that mud season commences. The beginning of mud season is earlier in southern New England, possibly starting as early as mid- to late March. More often than not, mud season begins up to one month after the spring equinox of late March. In northernmost New England, mud may not begin until May and extend into June. Of course, a warm dry winter will produce minimal mud and an earlier start to the season. However, for our purposes, we discuss weather and climatic conditions from mid- to late April and May into early June. This is the timing for "real spring" in New England. Initially, we discuss the controls and variability on temperature during this season, followed by a discussion on what controls precipitation variability across the region. We end with the several subseasons within mud season. One can probably define more subseasons at this time of year, given the quick transition toward summer, than in any other season we define.

Temperature

As daylight hours grow longer and the sun climbs higher in the sky in April and May compared to the dead of winter, temperatures begin to increase across the region (see fig. 2.4). The jet stream begins to move farther and farther north, allowing more air from subtropical regions to invade New England. At the same time, the strength of the large Canadian high pressure system that brings winter cold into the region is dissipating. Severe cold outbreaks in late April through May are not unheard of, but they are rare. At the same time, the Bermuda-Azores High is expanding in overall size, as well as moving further northward where its influence on New England may be easier felt. In addition, this high is strengthening. Temperatures do occasionally get into the 90s during mud season. Average temperatures in May range from near 60°F in northern New England to over 70°F in southern New England.

Freeze-thaw cycles culminate during the earliest part of the mud season (having reached their peak late in the ski season; fig. 8.1), which increases the expansion and contraction of soils beneath road beds (fig. 7.8). This phenomenon increases the occurrence of potholes every spring. Freeze-thaw cycles are most prevalent in the transition seasons, thus the process also is prevalent during autumn. However, the extreme cold, abundant moisture availability from snow melt, and the frozen ground just prior to mud season makes the effects of freezing and thawing much more pronounced at this time of year compared to the fall. The number of freeze-thaw days in central parts of the region exceeds 90, while fewer occur in northernmost New England and in coastal areas (fig. 8.2). Northernmost areas are consistently below freezing during more of the year compared to interior regions, thus the lower number of freeze-thaw days. Coastal regions are warmer overall, thereby reducing the number.

Fig. 8.1. Monthly distribution of freeze-thaw days for New England, showing that peak conditions usually occur over northern and interior sections of New England late in the ski season, although they can extend into the mud season. In some years, the results of freeze-thaw processes will be observed in mud season more than during the ski season. Modified from Schmidlin et al. (1987; fig. 3), *Journal of Applied Meteorology and Climatology*. Used with permission from the American Meteorological Society.

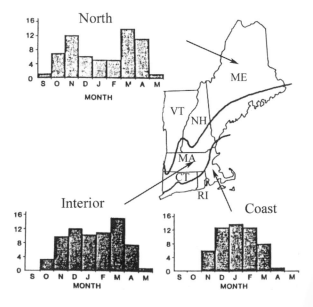

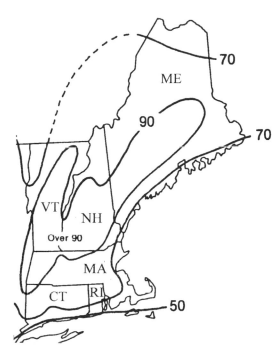

Fig. 8.2. Number of freeze-thaw days during the year across New England. Modified from Schmidlin et al. (1987; fig. 2), *Journal of Applied Meteorology and Climatology.* Used with permission from the American Meteorological Society.

Record low and high temperatures during mud season range from the teens to the mid-90s, respectively, in northern New England. For example, record May temperatures at Caribou, Maine, are 18°F and 96°F. In southern New England, record low and high temperatures are in the upper 20s to upper 90s. Hartford, Connecticut, record temperatures in May are 28°F and 97°F. Such a large difference in extremes is typical of the great amount of year-to-year variability that may occur within the transition seasons of New England.

Precipitation

The mud season is characterized by a very distinct transition in the nature of precipitation across New England. With the jet stream/polar front moving further north, storm tracks begin to shift (fig. 2.6). Storms moving up the Atlantic coast may still occur, but their frequency is beginning to decline, and, by the end of May, they are becoming more and more occasional. With this change in storm tracks, more warm-weather type storms are appearing. These are more local, convective storms—the type that produce brief showers and thunderstorms. However, peak thunderstorm season does not reach New England until late in the beach and lake season (chapter 9).

On the other hand, enough temperature contrast still exists between the ocean and land in the northeastern United States, and New England, in particular, to fuel development of coastal storms. Several major snow storms have occurred very late in April, and snow in May is not out of the question, especially in northern New England. In fact, one of the largest snow events in the seventy-plus years of record at Keene and at Hanover, New Hampshire, occurred on 12 to 13 April 1933. That storm dropped 23.0 inches of snow in Hanover and 21.0 inches in Keene. At the same time, nor'easters can produce significant amounts of rain, given the overall warmer temperatures of April and May, especially in coastal areas.

Precipitation averages for April and May range from over 8 inches atop Mount Washington in April and over 4 inches along the Maine coast in April to just over 2 inches at several locations in Vermont in May. The majority of this precipitation is in the form of rain. In fact, some large storms have produced over 5 inches (6.79 inches on 13 June 1998, Portsmouth, New Hampshire) to over 10 inches (6 June 1982, Cockaponset, Rhode Island) of rain during mud season, although both of these examples occurred very late in mud season. A potential problem that can occur in mud season is flooding caused by heavy spring rains on top of a deep snowpack at the beginning of the season. The most dramatic example of this scenario actually occurred near the end of the ski season, that is, the famous "All New England Flood" of March 1936. However, flooding is more likely to occur in the beginning of mud season, generated by a rain on deep snow event. In the case of the "All New England Flood," a deep snowpack existed early in 1936. Early spring warmth started the melt, but the speed of the melt became dangerous when steady rain began falling on 10 March. Flooding was present over the next few days, and when an ice dam broke on the Connecticut River, a major flood event followed. Another rainstorm produced record flooding by 18 to 19 March. These levels have not been exceeded, even by the 1938 hurricane (chapter 18). Luckily, the heavy snowpack of the 2000/2001 winter that covered much of New England (well over 4 feet in some places in early April) did not produce major flooding because the spring rains never came.

Subseasons

Of the major seasons in New England, mud season probably contains the greatest number of subseasons. We discuss four such subseasons: planting season, peeper season, the infamous black fly season, and the "fifth" season. All of these subseasons are highly dependent on the transition out of the ski season into the beach and lake season.

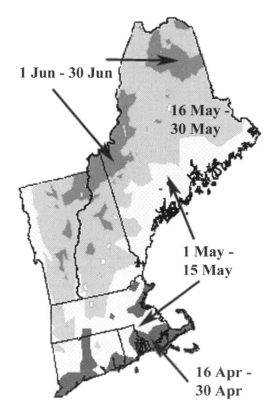

1 Jun - 30 Jun

16 May -
30 May

1 May -
15 May

16 Apr -
30 Apr

Fig. 8.3. Range of dates for the median last frost across New England. The last frost during mud season will occur within these timeframes 50 percent of the time. Modified from the *Climate Atlas of the United States* (National Oceanic and Atmospheric Administration/Department of Commerce, 2000; disk 1).

Planting Season. The traditional beginning of the growing season throughout much of New England is mid-May, with the traditional "planting of the home garden" occurring around Memorial Day. In part, this is a function of the average date of the last killing frost (fig. 8.3). That date varies from mid- to late April in southern New England to early June in northernmost New England. Last frost in most of northern New England is mid- to late May. An average planting date would be about 15 May. In addition to the absence of a killing frost by late May, the increased temperatures occurring during mud season start to warm up the ground, thereby allowing the planting of early crops. Peas are one of the earliest seeds to get into the ground. The flip side is that the start of planting season is highly influenced by mud season. In particular, heavy rains during April and early May would mean an extensive mud season and a longer wait for the ground to dry out to the point where it can be worked.

In addition to the impact mud season has on planting, weather conditions during this time of year can have a major impact on many crops. Apple trees cannot stand a frost or temperatures within a few degrees of 30°F during the bloom in mid-May. Furthermore, abundant rain during the bloom prevents honey bees and bumble bees from flying, thereby limiting pollination of the blossoms (according to apple growers Ernie and Shawn Roberts and Susan Jasse). Rain at night, however, will not affect the bees. On the other hand, cool temperatures (less than 50–55°F) and high winds (greater than 30 knots) also will prevent bees from flying (so say beekeepers Alden Marshall and Dick Dionne). Rain during the planting season, even if the ground is not muddy, still will delay planting of many vegetable crops, particularly corn. A late frost in early May can flush young growth of trees in tree farms. It may take one or two years to recover, and bugs may accompany the new growth (Northan Parr, personal communication).

Peeper Season. With the increase in temperatures through April into May, wildlife activity becomes much more intense. One of the more obvious signs is the mating calls of the small tree frogs, often referred to as spring peepers. Their wonderful sound during the breeding season often fills the evening and night air. Their sound truly characterizes the increasing warmth of the mud season as New England's climate makes the transition into summer.

Black Fly Season. In contrast to the wonderful sounds of the peepers, the mud season brings less welcome activity from another form of fauna. With the melting of snow and thawing of the soil comes increased flow in most streams throughout New England. Black flies lay their eggs in moving water. So, once the melt season is in full swing, the black flies begin to hatch (fig. 8.4). As a result, attacks on mammals (especially on humans) become more prevalent, and thus more annoying by late May to early June. The season only lasts for four to six weeks (extending into the beach and lake season), in most of New England. Black fly season begins and ends later in northernmost New England, extending into early July. Tolerance of, and even being proud of, black fly season epitomizes the heartiness of those living in the region. Black fly season also epitomizes the influence on humans of climatically controlled phenomena across the region. The amount of winter snow and springtime temperatures determine the timing and length of snow melt. The resulting runoff conditions highly influence the severity and length of black fly season.

"Fifth Season." In the March 2001 issue of *Yankee Magazine*, Castle Freeman, Jr., discussed a "fifth season" in addition to the traditional four. This season is very short, often occurring in March, that is, at the beginning of mud season. It is observed primarily in Vermont, but we are sure most New England states go through this "season." It comes between the late-season snows of ski season and the thaw of mud season, a time when New England's yearly climate is in transition. It is characterized by brown, tan, and yellow colors, pale and

Fig. 8.4. The dreaded black fly! Drawing from http://overton.tamu.edu/htmsub/blkfly3.html.

washed, thus it is not spring, and sugaring has ended. It is essentially an empty, in-between time.

Summary

Mud season means thawing. It is the time of year when temperatures are beginning to rise out of the cold of the ski season. However, major snow storms can occur during mud. Precipitation is similar to other times of the year, but that is the key to New England's hydrological regime. Freeze-thaw periods still can occur in mud season, and in some years, the bulk of freeze-thaw activity takes place during the typical months of mud. This season is highlighted by abundant subseasons, including planting season, peeper season, black fly season (a New Englander favorite), and the "fifth season," an empty nondescript period that marks the transition from ski season to mud season.

Beach and Lake Season

Snow, you can plow it back down to the pavement, salt it, or sand it.
But fog, you just have to accept it. —RICHARD ROONEY

Beach and lake season in New England is primarily limited to summer months. Summer in New England can be defined several different ways. First, there is the traditional "tourist season," which extends from Memorial Day through Labor Day. Prices at hotels and bed-and-breakfast inns (B&Bs) take a quantum step upward on Memorial Day, when tourist demand suddenly increases, then drop back down after Labor Day when the tourists quickly disappear and demand decreases. The waxing and waning of these prices are as predictable as the rising and setting sun. Other ways to define summer are thermal summer, which includes the ninety warmest days of the year on average. In New England, thermal summer usually begins about 10 June and runs through 10 September. Astronomical summer begins with the summer solstice (21 or 22 June) and ends on the autumnal equinox (near 22 or 23 September). Finally, climatic summer extends from 1 June through the end of August, and is frequently used for convenience by keeping whole months together in the same season. No matter how it is defined, summer represents the time of year to head to the water, or to higher elevations for relief from the New England heat.

In summer, days are long (fig. 2.4; table 2.1) and the sun's angles are high (fig. 2.3). The combination of these two factors leads to large amounts of energy being delivered to, and absorbed by, the New England landscape. In addition, the Atlantic Subtropical High, otherwise known as the Bermuda-Azores High, is usually at its maximum strength and size in summer, often driving warm moist air into New England from the south (figs. 2.6 and 2.7). Not only are daylight hours long and solar intensity high, but warmer and more humid air is being driven into our region by this quasi-stationary high pressure system over the northern Atlantic Ocean. As a result of these factors, summer temperatures here can get hotter than those found along the Gulf Coast of the United States. For example, the New England record high temperature of

107°F recorded at both Chester and New Bedford, Massachusetts, is warmer than the record high temperature recorded in both Miami and Orlando, Florida (table 9.1). The reason for lower maximum summer temperatures in these southern coastal climates is that when the temperature nears 100°F, there is almost always enough moisture available to produce a thunderstorm, which then cools off the surface. In New England, we occasionally experience high humidity, but summer afternoons can also be dry on a west wind. These dry west winds can push afternoon temperatures to the high 90s, but that doesn't happen as often as temperatures reaching into the 90s with high humidity.

Temperature

On average, afternoon high temperatures in New England typically range between 70°F and 80°F. During the hottest month, July, afternoon temperatures are closer to the mid-80s, while June and August are on the milder side, in the low 70s. However, note that temperatures in the 90s are common in southern New England. For example, Hartford, Connecticut, on average, experiences over fifteen days each summer with afternoon temperatures over 90°F (see more detailed discussion of extremes in chapter 11 and table 11.2). Northern New England, on the other hand, rarely will experience highs in the 90s. Table 9.1 shows the extreme high temperatures for each state in New England. Relief from these excessively warm days can sometimes be found along the beach, especially along the eastern coast of New England, where the cold water offshore often produces a notable sea breeze (fig. 9.1). When land in the coastal zone begins to warm in the afternoon, the colder, denser air from over the water surface rushes in and lifts the warmer, lighter air to higher altitudes in the atmosphere cooling coastal sites (fig. 2.2).

Table 9.1
Record high temperatures officially recorded
for each state in New England

State	Location	Date	Temperature (°F)
Massachusetts	Chester	2 Aug. 1975	107
	New Bedford	2 Aug. 1975	107
New Hampshire	Nashua	4 July 1911	106
Connecticut	Waterbury	22 July 1926	105
Vermont	Vernon	4 July 1911	105
Maine	North Bridgton	10 July 1911	105
Rhode Island	Providence	2 Aug. 1975	104

From Watson (1990, p. 171).

Fig. 9.1. Individuals enjoying a walk along a Cape Cod beach (*top*). The bluffs along many of the beaches on the outer Cape (*bottom*) add to their appeal, although these bluffs are susceptible to erosion from large waves associated with nor'easters and tropical systems. The sign evidences their fragility. Photos by Greg Zielinski.

Morning low temperatures tend to be in the low 50s in June and August, and in the upper 50s in July. Although at present the temperature almost never drops below freezing in summer at most of the larger metropolitan locations, it occasionally gets close. In the past few centuries, temperatures below freezing have been recorded in all three summer months (June, July, August). Of course, if you are located at higher elevations, all bets are off. Mount Washington has even been as cold as 8°F in the month of June. In general, the mountain summits can be quite a bit colder than valley locations. Temperatures will decrease at a rate of 3.5°F per 1,000 feet of elevation gain. This change in temperature is referred to as the environmental lapse rate. Many novice hikers have been caught unprepared for the chilly temperatures they faced in the White and Green mountains, and even the Berkshires. Many times, these chilly temperatures are compounded by high winds, hence an even lower wind chill temperature exists. Even in July, Mount Washington averages 53°F and 43°F for its daily high and low temperatures, respectively, and the warmest temperature ever recorded on the summit is only 72°F. We discuss Mount Washington in more detail in chapter 12. The point is, even if it is warm at the base of a mountain, do not be surprised at how quickly the temperature will drop and the winds increase on your way to higher altitudes.

Given these maxima and minima temperatures across the region, daily average temperatures in July vary from the low 70s along the southern coast to the low 60s in extreme Down East Maine (fig. 9.2). July means in the mid-60s are also found in northern Maine and toward the central hilly region of Vermont and New Hampshire. Overall, July average temperatures may be expected to be lower along the coast because of the frequent occurrence of a sea breeze.

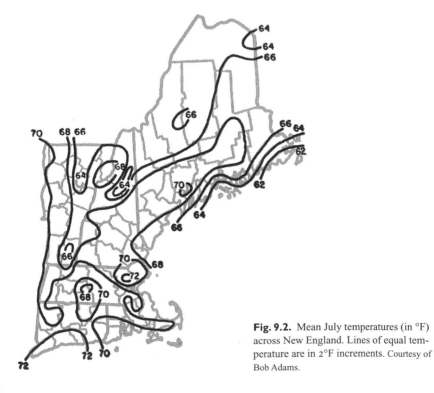

Fig. 9.2. Mean July temperatures (in °F) across New England. Lines of equal temperature are in 2°F increments. Courtesy of Bob Adams.

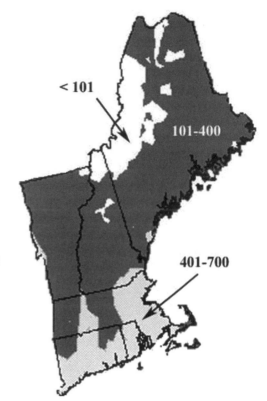

Fig. 9.3. Median number of cooling degree days (CDD) across New England. Numbers represent the number of CDD during the cooling season one would expect to receive 50 percent of the time. Calculation of CDD is described in the text. Modified from the *Climate Atlas of the United States* (National Oceanic and Atmospheric Administration/Department of Commerce, 2000; disk 1).

However, the prevailing south to southwesterly flow over the region during the summer strongly influences temperatures on the south-facing coast of mid-coast to Down East Maine.

Visitors to New England during the summer should be mentally prepared for the lack of air-conditioning. Afternoon temperatures are typically mild enough to endure without this modern convenience. However, there are a few days every summer when everyone wishes they were cooler. It just does not make sense in most cases to make this investment, when the need for such an item is so short-lived. One measure of summer comfort, or lack thereof, is the number of cooling degree days (CDDs). CDDs are not really days, but are units of energy that are needed to reduce the effective temperature to a comfortable level. The formula assumes that some cooling would be required anytime the daily average temperature is above a base of 65°F; hence you accumulate one CDD for every degree in excess of the base. New England rarely exceeds 400 CDDs per summer, and the measure is commonly between 250 and 350 (fig. 9.3). To put this in perspective, Miami, Florida, averages 4,198 CDDs per year. In New

England, heating in winter takes precedence over cooling in summer, because the Heating Degree Days (HDDs), based on average daily temperatures below 65°F, average about 5,000 in southern New England and over 9,000 in northern New England (see chapter 7, fig. 7.3). As one would expect, Miami accumulates only about 200 HDDs. Consequently, most New Englanders invest in good heating systems, and tough out the unpleasant summer heat with fans, open windows, and cool beverages. Of course, a trip to the beach or into the mountains is an added deterrent to the purchase of an air conditioner.

Precipitation

In summer, rainfall reaches its yearly minimum along New England's eastern coast because of the impact of the sea breeze on thunderstorm activity; that is, the cool sea breeze mitigates the formation of afternoon thundershowers. Interestingly, the eastern coast of New England is the only region in the eastern United States that has a cold water current offshore, and as a result, this coastal zone averages fewer thunderstorms per year than anywhere else east of the Rocky Mountains (fig. 5.2). On average, the number of thunderstorm days in New England is twenty or fewer, whereas Pennsylvania often experiences thirty to forty thunderstorm days. Florida will have up to one hundred thunderstorm days per year, the highest rate in the contiguous United States. Although they are even fewer and farther between, some New England thunderstorms have produced tornadoes (fig. 9.4). A small tornado alley has been documented

Fig. 9.4. Thunderstorm over Bangor, Maine, with a reasonably well-developed anvil top, particularly for a New England thunderstorm. Photo by Greg Zielinski.

in western Massachusetts, though even here, tornadoes are nowhere near as prevalent as on the Great Plains of the United States (Leathers, 1994). Nevertheless, the most classic example of the devastating nature of New England tornadoes is the Worcester tornado that touched down in the late afternoon in June 1953, claiming ninety lives over its path. We discuss tornadic activity in more detail in chapter 17.

However, the sea breeze effect does not penetrate very far inland. For example, there is no evidence that Concord, New Hampshire, which is less than 50 miles from the coast, has a reduction in summer precipitation like that found at Portland, Maine, and Boston. Furthermore, locations farther inland, including Burlington, Vermont and Caribou, Maine, tend to have their wettest months overall in summer. This is discussed in more detail in chapter 11 (see fig. 11.5). In addition, higher elevations receive significantly more precipitation in all seasons, including summer, when thunderstorm activity is at its peak with the rising of heated air up the mountains. Mount Washington averages over 23 inches of liquid equivalent precipitation in June, July, and August. In contrast, Pinkham Notch, in the valley below Mount Washington, averages less than 15 inches for the same period, and North Conway, just a hop, skip, and jump from Pinkham Notch, receives less than 12 inches. The increase in rainfall by elevation is well documented and is caused by the uplift of traveling air masses. The air lifted by the mountain cools through expansion as atmospheric pressure decreases, causing condensation and precipitation. Details on the processes active in alpine environments are in chapter 12.

Subseasons

Black Fly Season. Black flies tend to serve as a nuisance in early summer, particularly in northern New England. As we explained in chapter 8, black flies need moving water to reproduce. In late spring and early summer, snowmelt in the mountains is plentiful and soils are mostly saturated with water. However, as temperatures climb into mid-summer, water supplies tend to diminish and black flies slowly fade away. They disappear at lower elevations first, usually by mid-June (that is, just after the end of mud season). On the other hand, they persist quite a bit longer at higher elevations with their lower overall temperatures, babbling brooks, and higher moisture availability. This is also true for more northern locations. The cool summer of 2000 enabled black flies to persist for the entire summer in parts of Maine! These pests can be quite abundant and annoying at the peak of their season, and the bites of the females leave behind large itchy welts that usually take days to subside. These critters are so problematic during certain periods that many folks in New England plan their vacations around "black fly season" to lessen their detrimental impacts on the joys of sun, sand, and hiking. Luckily, in most years it is just the beginning of summer that falls into black fly season.

Growing Season. The growing season of summer is a time period when specific weather and climatic conditions may have a significant impact on the harvest subseason discussed in chapter 10. For instance, too much heat and humidity can cause apples to drop and lead to scab or fungus disease (Ernie and Shawn Roberts and Susan Jasse, personal communication). Drought conditions during the summer can lead to browning of trees in tree farms (Northam Parr, personal communication). Very hot summers also can have an impact on livestock. Interestingly, hot summers can warm ocean temperatures to the point that lobsters will move close to shore, shed, and produce soft shells (Carl and Carol Ann Widen, personal communication). These lobsters are not good to eat. The heavy fog that can be produced during beach and lake season also makes it difficult for lobster fisherman to troll and recover their traps without running over other fishermen's lines.

Summary

The beach and lake season is the hottest part of the year in New England, but temperatures are relatively cool, on average, compared to many parts of the country. Highest temperatures are found in southern New England, although the development of a sea breeze will cool the southern shores. High temperatures in northern Maine remain in the 70s on average. Thunderstorm activity reaches it peak this time of year, but New England only has about twenty thunderstorm days per year on average. This is the lowest number east of the Rocky Mountains. The develoment of a sea breeze along the coast stabilizes the air and prevents thunderstorm growth in those areas. Interior-most parts of the region do have more thunderstorms and as a result, they can have a peak in precipitation during the summer. Coastal areas often have a minimum in precipitation during the summer.

Chapter 10

Foliage Season

Agreable harvest weather this month, and a verry good Crop of Corn in general,
both English and indian, also potatoes, tirnips, and all the fruits of the Earth
this year as ever I knew. —SAMUEL LANE, STRATHAM, NEW HAMPSHIRE,

OCTOBER 1795

Foliage season provides yet another opportunity, not only for tourists, but residents as well, to partake in the wonderous beauty of New England. The wonderful colors of the foliage season, like many other New England splendors, are highly influenced by the weather and climate of the region. Depending on location and tree species, foliage season spans all three autumn months—September, October, and November—though peak color conditions are usually best in October. Late November is associated with the beginning of ski season (chapter 7). Interestingly, small splashes of color may be seen as early as late August, a great surprise to those who live further south in the United States. However, this phenomenon is probably caused by some other type of environmental stress to specific trees, as opposed to the climatic changes close to the start of the foliage season.

Many locals argue that foliage season is the most beautiful time in the region. Compared to other locations in the Appalachians, New England is unique because it contains an assemblage of both hardwoods and softwoods, including sugar, red, and swamp maple, white pine, yellow and white birch, white and red spruce, red and white oak, balsam fir, and hemlock, as well as a variety of deciduous shrubs. This variety of trees produces a splendid patchwork of colors, including brown, green, yellow, orange, and red. It is primarily the abundance of various maples, which produce bright red leaves before defoliating, that distinguishes New England from its foliage-season competitors to the south. Although the peak of the season generally occurs in October, peak conditions do not happen region-wide at the same time, but rather migrate from north to south (fig. 10.1). This migration is related to temperature and sun angles, which in turn control photosynthesis. It should also be noted that many other variables influence the timing and intensity of the foliage season, including the temperaure regime leading into the season, soil moisture conditions especially as influenced by summer precipitation, and cloudiness. Each tree species is affected in a slightly different way. Once a tree ceases to photosynthesize, it also stops producing chlorophyll, the substance that turns the leaves green. Although much debated, the best conditions for vibrant foliage colors are warm sunny days and cool nights, not the presence of a hard frost. The reddish leaves are produced chemically because of the cooler temperatures, while the yellow and orange pigments in the birch and maple trees are always present in the leaves, but are masked by the chlorophyll. Note that the coastal zone tends to peak a little later than areas farther inland at the same latitude (fig. 10.1). This is related to ocean temperature, which tends to change very slowly. Therefore, as we march into the foliage season, ocean temperatures tend to have a warming influence in the coastal zone, which delays the onset of the season.

Temperature

During the fall season, the length of daylight hours declines rapidly, at a rate of nearly 3 minutes per day in late September and early October (fig. 2.4). The

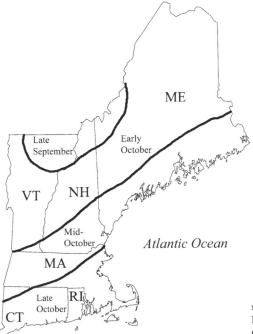

Fig. 10.1. Most frequently observed timing of peak colors across New England. Adapted from Weather Channel home page, www.weather.com.

sun's intensity is also lessened as its elevation above the horizon at noon becomes lower and lower (fig. 2.3). As a result, air temperatures decline quickly. For example, at Durham, New Hampshire, average maximum and minimum temperatures on 1 September are 79.4°F and 52.1°, respectively. By 30 November, these values drop to 44°F and 25°F, respectively—approximately 30°F colder in two months. The change in temperatures from the beginning of this season to the end are even more obvious in northern New England. The average daily high in Caribou, Maine, is 68°F on 1 September while the daily low averages 47°F. This is in stark contrast to these same values on 30 November. By the end of November, average high and low temperatures in Caribou are below freezing: 30°F for a high and 14°F for a low. In southernmost New England, Hartford, Connecticut, goes from an average daily high of 78°F and low of 56°F on 1 September to 44°F for an average daily high and 27°F daily low on 30 November. From the beginning to the end of "climatic autumn," the region shifts from beach mode to winter mode, with foliage season sandwiched in between.

Interestingly, the foliage season may contain a climatic singularity, defined as an event that is persistent in its timing from year to year. The January thaw is the other New England climate singularity (chapter 7). The specific phenomenon that commonly occurs in the fall, although its timing is not regular enough for it to be considered a true climatic singularity, is Indian Summer. Indian Summer occurs when temperatures rise well above average following the first killing frost. The incursion of warm air into New England during autumn is a result of high pressure systems migrating to the south and southeast of the region. As the jet stream has yet to be found consistently south of New England, sub-tropical air can easily find its way back into New England during foliage season. In fact, alternating incursions of polar air and subtropical air into New England is characteristic of autumn in the region.

Precipitation

In many central and coastal sites around the region, fall is the wettest time of the year. July and August are usually the wettest months in the northernmost interior parts of New England. For instance, there is a general increase in precipitation from September to November in the southeastern part of the region and along the entire coast, with November being the wettest month at most locations. The northwestern locations in the region tend to have less rain in October than September. A considerable proportion of average September rain is produced by tropical weather systems, such as hurricanes and tropical storms that make landfall either in New England or elsewhere in the United States. The region experiences a landfalling hurricane or tropical storm that maintains its tropical characteristics roughly every five years, on average. However, some tropical systems become more extratropical as far as their structure by the time they reach New England, while many tropical systems arrive via land routes after making landfall elsewhere in the eastern United States. A detailed discussion of hurricanes is in chapter 18.

By November, nor'easters begin to increase in frequency. Although the greatest number of nor'easters occurs later in the winter season, some of the most intense nor'easters may occur in November, when sea surface temperatures are still relatively warm and the jet stream is more frequently positioned over New England. As a result, these storm systems can carry large amounts of moisture and can produce very heavy rainfall at this time of year. Significant snowstorms are also common in November. Northern New England experienced an impressive snowstorm as recently as 14 November 1997, with a combination of sleet and freezing rain reported in southern New England. The vicinity of Portland, Maine, recorded 15 to 20 inches of snow during this November event. This is but one of many early season snowstorms that have affected the region.

In fact, one of the most notorious storms to hit New England occurred during the foliage season. This storm exemplifies a key aspect of storm systems that may impact the region, that is, the potential interaction between tropical

Fig. 10.2. Satellite imagery (GOES-7) of the "All Hallows Eve" or "The Perfect Storm" on 30 October 1991, at 1201 UTC. The original photo was infrared color enhanced. Photo from the NOAA/National Climatic Data Center (National Oceanic and Atmospheric Administration/Department of Commerce) web site at http://www.ncdc.noaa.gov/pub/data/images/extratrop-atlantic-19911029-g7vis.jpg.

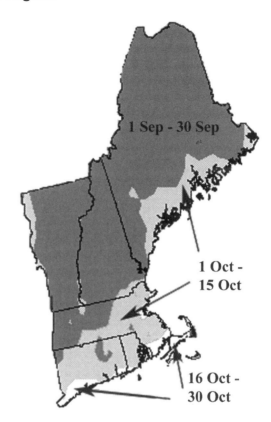

1 Sep - 30 Sep

1 Oct - 15 Oct

16 Oct - 30 Oct

Fig. 10.3. Range of dates for the median first frost across New England. The first frost during foliage season will occur within these timeframes 50 percent of the time. Modified from the *Climate Atlas of the United States* (National Oceanic and Atmospheric Administration/Department of Commerce, 2000; disk 1).

and extratropical systems. Now known as "The Perfect Storm" because of Sebastian Junger's book and the subsequent movie of that title, the "All Hallows Eve" storm of 1991 was the product of a very strong nor'easter that eventually merged with and drew energy from Hurricane Grace (fig. 10.2). The interesting aspect of the storm was that it initially formed off Sable Island, just south of Newfoundland. The storm then retrograded, that is, it backed to the southwest, coming closer to the New England coast, then eventually closer to the Middle Atlantic coast. The greatest impact of the storm was the exceptionally heavy surf that caused considerable damage along the coast, beach erosion, and flooding, especially during periods of high tide.

Subseasons

Harvest Season. The first freeze after the growing season also occurs sometime in autumn with the harvest occurring before the "frost is on the pumpkins"—if all goes well (figs. 10.3 and 10.4). Growing seasons in New England are relatively short, with the growing season being defined as the length of time between the last spring freeze and the first autumn freeze. The growing season persists for approximately 180 days along the south shores of Connecticut and Rhode Island and drops down to 120 days or fewer in northern Maine, New Hampshire, and Vermont (fig. 10.5). Growing seasons are even shorter at higher elevations. However, all crops (or natural vegetation) are not decimated by a mere light freeze. To end the growing season, crops must experience a killing frost; the severity of cold varies somewhat between species. A frost will

Fig. 10.4. Two New England favorites at harvest time, pumpkins and apples from West Meadows Farm in the Champlain Valley of Vermont. Photo by Jim Gallot, West Meadows Farm, New Haven, Vermont, as seen at http://gonewengland.about.com/travel/gonewengland/gi/dynamic/offsite.htm?site=http://www.westmeadowsfarm.com/.

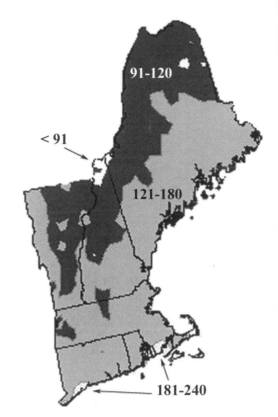

Fig. 10.5. Average length of growing season across New England, defined as the number of days between the last killing freeze in spring and the first killing freeze in fall. Modified from the *Climate Atlas of the United States* (National Oceanic and Atmospheric Administration/Department of Commerce, 2000; disk 1).

occur when the temperature falls to 32°F, but the actively growing tissues of most plants are not injured until temperatures drop to 23° to 30°F, that is, a killing freeze (Moran and Morgan, 1997). In fact, the growing season in parts of New England is among the shortest in the country. Nevertheless, the extended summer daylight hours lead to relatively productive agricultural endeavors in a short period of time, because more hours are available for photosynthesis.

Summary

Autumn in New England brings a wide variety of weather conditions to the region. It begins with summer-like air temperatures and very warm lake and ocean temperatures. It ends with the prospect of paralyzing snow storms and frigid temperatures. Nestled between these highly contrasting climate conditions are the beautiful orange, red, green, and yellow leaves that are responsible for what we call the foliage season. The decreasing amount of sunlight, greater differences in temperature between warm days and cool nights compared to summer conditions, and overall cooler temperatures are responsble for the enjoyable leaf-peeping conditions. On the other hand, the occasional landfalling hurricane or tropical storm and increasing occurrence of nor'easters produce the wettest time of the year for much of the eastern and coastal sections of New England. The potential interaction between these two types of storm systems may make for some interesting conditions, including heavy rainfall, strong winds, and high surf. "The Perfect Storm" exemplifies such a scenario.

Chapter 11

The Year in Summary

He doesn't know what the weather is going to be in New England . . .
probable northeast to southeast winds, varying to the southward and
westward and eastward, and points between, high and low barometer
swapping around from place to place; probable areas of rain, snow,
hail, and drought . . . —MARK TWAIN

The season-to-season variability in New England's weather and climate can be summarized by identifying mean annual temperature (MAT; fig. 11.1) and mean annual precipitation (MAP; figs. 11.2 and 11.3) across the region. This provides excellent insight into the spatial changes across the region and how the different climatic controls discussed earlier in the book are manifested in actual numbers. Mean annual numbers also are very easy to use to determine how New England's climate has changed over time, whether any repetitive cycles can be seen in these changes, and if any distinct trends occur in annual values (such as an increase in annual temperatures over the region). We discuss these details in chapter 19. This information is then used to evaluate—and ideally predict—what may happen to New England's climate in the future, as in chapter 20.

However, we must keep in mind that year-to-year variability can be quite striking when we evaluate MAT and MAP. We give an example of this variability by presenting the standard deviations (SD) and extreme values for mean annual temperature, precipitation, and snowfall for Durham, New Hampshire (table 11.1). One interesting aspect of these three parameters is the great variability that can occur in precipitation compared to temperature averages and extremes: The SD for mean annual temperature in Durham is 3 percent compared to 18 percent for mean annual precipitation. The SD for mean annual snow is much greater, a whopping 37 percent. The difference between the highest and lowest recorded temperature and precipitation likewise reflects how precipitation can be so much more variable from year to year than can temperature. Actually, the great year-to-year variability in precipitation is a long-term reflection of the great variability in precipitation amounts that can occur within a single storm.

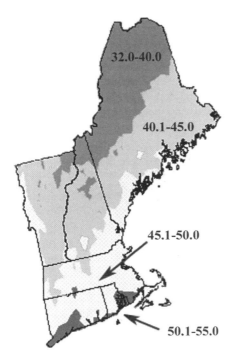

32.0-40.0

40.1-45.0

45.1-50.0

50.1-55.0

Fig. 11.1. Mean annual temperature (MAT) in °F, across New England. Modified from the *Climate Atlas of the United States* (National Oceanic and Atmospheric Administration/Department of Commerce, 2000; disk 1).

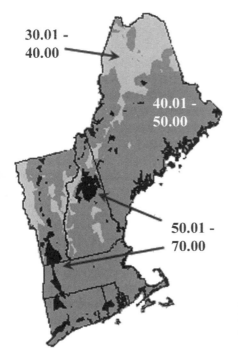

30.01 - 40.00

40.01 - 50.00

50.01 - 70.00

Fig. 11.2. Mean annual precipitation (MAP) in inches, across New England. Modified from the *Climate Atlas of the United States* (National Oceanic and Atmospheric Administration/Department of Commerce, 2000; disk 1).

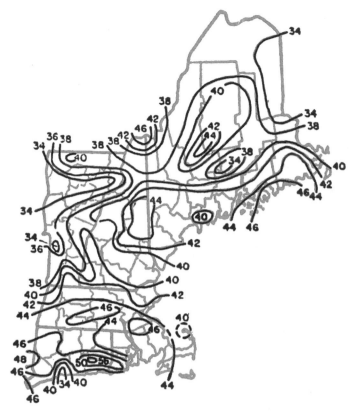

Fig. 11.3. Detailed mean annual precipitation (MAP) inches, across New England. Higher values associated with mountainous areas of interior New England are not shown, nor are the high values found in places along the coast of Maine. Lines of equal precipitation are in 2-inch increments. Courtesy of Bob Adams.

Moreover, depending on which years are used to determine average values, different means may be attained. Climatologists generally use thirty-year normals, or averages, to characterize any climatic shifts at a site. Thus, normal or average values change with time. We look at long-term trends in MAT and MAP in chapter 19. Also, it is important to remember that overall regional patterns across New England may not vary to a large extent with time, but specific, detailed patterns over smaller geographical sections may. The MAT and MAP values used in the following discussions are average values for the period 1961 to 1990. We also discuss several other aspects of the annual climate, including the prevalence of fog, the seasonality of temperature and precipitation, and extremes.

Table 11.1
Variability in average climatic values
for Durham, New Hampshire, 1895–1994

	Temperature (°F)	Precipitation (inches)	Snow (inches)
Mean	46.7	41.0	55.0
Standard deviation (S.D.)	1.3	7.5 (18%)	20.6 (37%)
Normal range	45.4–48.0	33.5–48.5	34.4–75.6
Minimum as recorded	42.9	24.0	13.5
Maximum as recorded	50.2	60.2	106.7

Standard deviation is a measure of the amount of variability in the record about the mean. For example, the annual average temperature for Durham is 46.7°F. Given the standard deviation of 1.3°F, the annual average temperature for Durham will be 46.7 ± 1.3°F, or it will fall between 45.4°F and 48.0°F 66 percent of the time, or two out of every three years. The mean plus or minus the standard deviation yields the normal range values in the table.

Temperature

Mean annual temperature (MAT) primarily varies as a function of latitude, elevation, and proximity to the ocean. New England's highest MATs are located in southern Connecticut and Rhode Island with another area of higher annual temperatures in southern Massachusetts (fig. 11.1). MATs in these areas are generally around 50°F. The decrease in MAT is generally in a northern direction, although it is more northwesterly from the Maine coast to northernmost Maine, essentially paralleling the coastline. Coldest MATs are found in northernmost Maine and New Hampshire, northeast Vermont, and in the hills of central New England. MAT in northernmost New England is generally around 40°F, as is the case for Pittsburg, New Hampshire, and Caribou, Maine. Intermediate MATs, around 45°F, are found in southern Vermont, New Hampshire, and Maine. Also, note that Mount Washington's MAT is near 26°F, the ultimate New England example of how elevation affects temperature.

Precipitation

Spatial trends in mean annual precipitation (MAP) across New England follow a much more complex pattern than those for MAT (figs. 11.2 and 11.3). MAP includes the conversion of snowfall totals to amount of water—that is, the snow water equivalent or s.w.e. Although latitude, elevation, and continentality (that is, the degree of continental influence on a site's climate as opposed to maritime or oceanic influence) play significant roles in determining MAT, storm tracks and topography are the dominant factors influencing MAP. For instance, the complex spatial changes in MAP across New England may be mirrored in the complex pattern of precipitation from a single storm (see fig. 14.9 for the October 1996 storm).

Overall, MAP amounts range from minimum values around 35 inches per year in northwestern Vermont and northernmost Maine to 50 inches or more per year in southern Connecticut and some places along Mid-Coast and Down East Maine. Mountainous terrain generally receives greater amounts of precipitation than lower elevations, everything else being equal (such as latitude). Mount Washington receives 99 inches of liquid equivalent precipitation per year (details about MAT and MAP in alpine regions is discussed in chapter 12). Consequently, greatest MAP values are found fairly close to the coast, as expected given proximity to moisture sources, and on the higher mountains. Orographic uplift greatly enhances precipitation values for a single storm (see discussion in chapter 12 and fig. 15.9 as an example of enhanced snowfall in the mountains of New Hampshire and Maine). This is especially true on the eastern side of the mountains as flow off the ocean from large coastal storms can produce heavy precipitation amounts on the windward side of the hills. In contrast, MAP values on the northwest side of the interior mountains are generally lower given the possible rainshadow effect from coastal storms. This held true for the 22 to 28 February 1969 storm (fig. 15.9).

Another area in New England has precipitation values that are higher overall than the surrounding areas, despite being located more inland. The enhanced precipitation of northern Vermont, New Hampshire, and northwest Maine may be a reflection of storms moving northeastward through the Saint Lawrence River valley (fig. 11.3). Such storms may produce significant precipitation in these areas, but would not produce as much precipitation south and east of the mountains, and particularly toward the coastal plain. In this case, the rainshadow is east of the central highlands. The high precipitation values in northern New England also could be a reflection of higher winter snowfall with storms tracking through the Saint Lawrence River valley, storms that may not be large enough to impact southern New England. Interestingly, winter precipitation is equal to or exceeds summer precipitation in only two regions east of the Rocky Mountains in the United States (Trewartha, 1981). Easternmost New England is one of those places. The other area is downwind of the Great Lakes, given the propensity for lake-effect snows. The high amount of winter precipitation in New England relative to summer precipitation is, in part, a function of the converging storm tracks over the region (with more storms occuring in winter than summer), as well as the lack of abundant summer thunderstorms in New England (as explained in chapter 9).

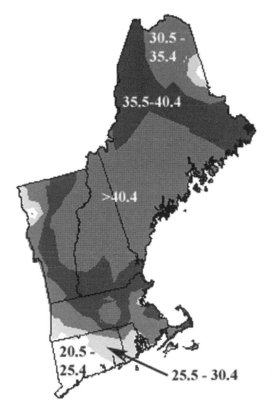

Fig. 11.4. Number of days per year that heavy fog, defined as visibility of 0.25 mile or less, occurs in New England. Modified from the *Climate Atlas of the United States* (National Oceanic and Atmospheric Administration/ Department of Commerce, 2000; disk 1).

Fog

Several other interesting meteorological phenomena play important roles in the region's climate. One of these is fog. The amount of fog recorded in New England is among the highest in the country. In particular, east-central New England (New Hampshire and central Maine) to the coast of Maine experiences heavy fog, defined as visibility less than 0.25 miles, on more than 40 days per year (fig. 11.4). The only other parts of the country with an equal number of heavy fog days are the southern spine of the Appalachian Mountains (parts of West Virginia, western Virginia and North Carolina, and eastern Kentucky and Tennessee) and in the Pacific Northwest (the Olympia Peninsula and southern Puget Sound, Washington, and coastal Oregon). The number of heavy fog days decreases rapidly (that is, basically halves) toward the corners of New England. As we explained in chapter 9, the greatest number of foggy days occur in the summer, when the contrast between the heated land and cool ocean waters is at a maximum. As expected, coastal areas are most susceptible to these con-

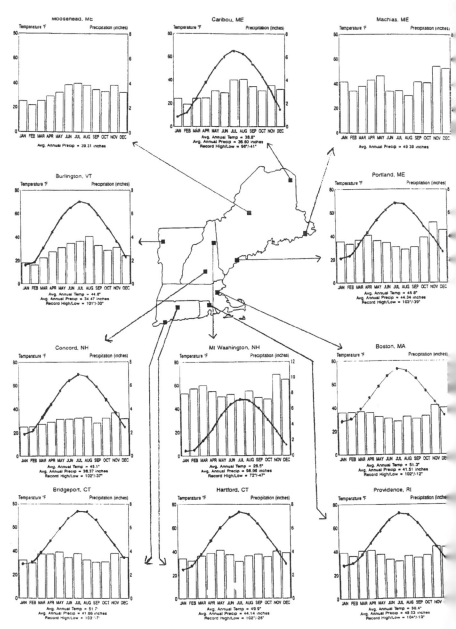

Fig. 11.5. Mean monthly temperature and precipitation through the year for several stations across New England as a means of showing seasonality across the region.

ditions; almost the entire eastern coast of New England has more than 35.5 days with heavy fog. This contrasts with the southern New England coast, where fewer than 30.5 days with heavy fog are counted in the average year. These numbers reflect the impact of the cool Labrador Current that flows into the Gulf of Maine compared to that of the warmer Gulf Stream that flows off the southern New England coast (see chapter 5; fig. 5.2).

Seasonality

Although we can summarize mean annual temperature and precipitation variability across the region, for many purposes, the difference between the maximum and minimum temperature and the timing of the maximum and minimum precipitation may be more important than the annual numbers. We summarize monthly temperature and precipitation for the year for several sites across the region to emphasize the seasonality of the region (fig. 11.5). From a temperature perspective, inland sites such as Caribou and Burlington experience greater seasonality, as we have discussed. Without the moderating effects of the ocean, these sites are much more continental, thus summer highs are greater than those for sites along the coast at the same latitude and winter minima are lower than those for sites at the same latitude along the coast. Note the much lower temperatures for Mount Washington, as discussed in detail in chapter 12 (fig. 11.5).

Seasonality in precipitation is a bit of a misnomer from an annual perspective. On a month-to-month basis, precipitation amounts often are between 3 and 4 inches regardless of where in the region one lives. There are some exceptions, such as the low precipitation amounts for February in Burlington, but from a region-wide perspective, the area is essentially "wet" every month. Wet conditions prevail each month partly because almost all storm tracks converge on New England, and because, even in the summer, the jet stream and its accompanying major storm tracks lie close enough that trailing frontal systems can still bring showers to New England. The convergence of storms in the region, especially during the winter, essentially provides a continual source of potential precipitation events throughout the year.

However, seasonality can be noted as far as the timing of the maxima and minima precipitation around New England, despite the fact that the difference in precipitation between the wettest and driest month may only be about an inch. A particular month or months may be a bit wetter or drier than others because of continental versus maritime controls. The most interior portions of the region—northern Maine and Vermont—receive most of their precipitation in summer, because they have more afternoon showers or thunderstorms during the summer than coastal areas (fig. 11.5). As we mentioned in chapter 9, the cool ocean waters inhibit thunderstorm growth closer to the coast. In fact, coastal sites generally receive the least precipitation during the summer for this very reason. Summer precipitation in northern interior sites also may occur because of trailing frontal systems brushing across that area as the main storm center moves well north of New England. These northernmost regions have their minimum in precipitation in February. The dominance of the cold Canadian High in that month may have a tendency to push storms further south in New England and the mid-Atlantic states. Most central and coastal sites often receive their greatest amount of precipitation in November, perhaps reflecting the yearly resurgence of the coastal storm track during the fall and winter seasons. The southerly migration of the jet stream and more frequent storms through the Saint Lawrence River valley in November may also contribute to this November maximum in precipitation.

Daily Extremes

Changes in extreme temperature and precipitation limits per year are summarized in table 11.2 (note that these numbers would change when considering alpine sites). As expected, the number of days that 90°F is equalled or exceeded changes dramatically going from north to south. Most areas across the northern tier of New England will have fewer than four days per year where the high temperature exceeds 90°F. Stations at slightly higher elevations in southern Vermont and New Hampshire also may have a very low number of days with such heat, as will sites in the hills of central Massachusetts (such as Worcester). The highest number of days per year recording greater than or equal to 90°F occurs in central Connecticut, as exemplified by Hartford's average of eighteen per year. Similarly, the number of days that the high fails to get above freezing changes dramatically from 80 to 90 or more days per year in northernmost New England, to 50- to 60-plus in central New England, to fewer than 35 in southern New England. The overall trend is the same for daily lows, as more than half of the year will have a low temperature below freezing in the north and less than 100 days will be below freezing in the south and along the southeastern coast. Rarely will overnight lows get below 0°F in southernmost New England. Over 30 to 40 days per year will be below 0°F in northern New England.

Surprisingly, very little difference can be seen across the region as far as the number of days when at least 0.1 inches of precipitation is recorded. For the most part, the entire region will have 75 to 85 days with precipitation at least 0.1 inch. On the other hand, days with heavier precipitation (more than 1 inch) are more frequent in southern and coastal New England, as expected given greater proximity to the primary moisture source, that is, the Atlantic Ocean. Many sites close to the coast or within about 50 to 60 miles of the coast (such

Table 11.2

Average number of days per year that specific extreme high and low temperatures and specific precipitation thresholds are exceeded or equalled for various cities across New England, 1970–1999

Station	Maximum temperatures $\geq 90°F$	$\leq 32°F$	Minimum temperatures $\leq 32°F$	$<0°F$	Daily precipitation >0.1 in.	≥ 1 in.
MAINE						
Bangor	4	63	154	20	81	11
Caribou	1	96	183	45	87	3
Portland	4	45	150	10	77	12
Rangeley	0	85	202	55	85	6
NEW HAMPSHIRE						
Berlin	4	68	181	35	85	9
Concord	11	50	172	20	77	11
Keene	10	42	167	18	81	10
VERMONT						
Burlington	6	66	151	24	81	5
Rutland	4	50	157	18	82	5
St. Johnsbury	7	60	167	30	85	6
MASSACHUSETTS						
Amherst	12	30	155	12	80	13
Boston	15	25	94	0	77	12
Worcester	3	49	139	6	81	13
RHODE ISLAND						
Kingston	5	20	135	3	79	13
Providence	11	24	115	2	79	12
CONNECTICUT						
Bridgeport	6	22	97	0	77	12
Hartford	18	33	130	5	81	12

From Garoogian (2000), after data from the National Climatic Data Center (National Oceanic and Atmospheric Administration/Department of Commerce). Used with permission from Grey House Publishing, Inc.

as, Concord, New Hampshire) will have around 10 to 13 days per year with more than 1 inch of precipitation. Interior sites are more likely to have 5 to 6 days with precipitation of this magnitude, and much of that could be snowfall (using the general water equivalent of 10 inches of snow equalling 1 inch of liquid precipitation).

Summary

Mean annual temperature (MAT) varies across New England from 40 to 50°F going from north to south and from the interior to the coastline. These numbers do not include alpine areas. Coldest locations are northernmost Maine, whereas coastal Connecticut, Rhode Island, and Massachusetts are the warmest. Mean annual precipitation (MAP) varies from about 35 inches in the more interior part of the region to over 50 inches along some coastal areas. The central and western highlands can receive over 50 inches of precipitation per year. In addition to these values, central New England to the coast of Maine will have over 40 days per year on average with heavy fog (visibility less than 0.25 mile). This is one of the foggiest areas in the continental United States. Greater seasonality of temperature (that is, higher summer and lower winter temperatures) are a function of the degree of continentality of the climate. A greater annual range occurs in the interior-most parts of the region. The same holds true for precipitation. Although most of the region will see monthly precipitation totals around 3 to 4 inches per month throughout the year, there are months with slightly more or less precipitation. Interior sections, particularly the northern interior, will have a summer maximum in precipitation, probably due to increased shower and thunderstorm activity compared to coastal regions. Summer minima in precipitation occurs in many coastal sites. Many coastal and central interior sections of the region experience a November maximum in precipitation, possibly in part from the greater frequency of coastal storms in the fall.

The Alpine Zone

We had recorded the greatest wind ever measured on earth.
—WENDELL STEPHENSON, 12 APRIL 1934

Mountain weather is notoriously unpredictable. Furthermore, it can differ markedly from the weather conditions only a short distance away at lower elevations on the valley floors (fig. 12.1). Anyone who has taken a drive up the Mount Washington auto road has experienced the changing climatic conditions with increasing elevation. Generally, temperatures decline at a rate near 3.5°F per 1,000 feet of elevation. This change of temperature with elevation is referred to as the environmental lapse rate. However, a rising parcel of air cools at the dry adiabatic rate of 5.5°F per 1,000 feet, at least until the parcel reaches saturation or 100 percent relative humidity. Heat is released when water vapor condenses, thus the rate of cooling is less when condensation occurs (that is, cloud formation occurs). On the other hand, several parameters, such as precipitation and wind speeds, tend to increase with height (less land surface to slow air movement).

As a result, vegetation gradually changes upslope to species more tolerant of harsher conditions than in the valley bottoms (fig. 12.2). On Mount Washington, for example, northern hardwoods dominate the base of the mountain up to about 3,000 feet. Tree species here include beech, yellow and white birch, sugar maple, and several smaller understory species such as moose maple, cherry, and hobble bush. Between 3,000 and 4,500 feet, the climate shifts into the boreal zone, which is made up of softwoods, including red spruce and balsam fir, with an occasional mountain ash or white birch. At about 4,500 to 5,000 feet a transition to the alpine zone occurs, where the species are the same as in the boreal zone, but they become dwarfed by the extreme windy and cold weather and the short growing season, producing Krummholz, or "crooked wood." The true "above treeline" areas are sparse in vegetation, but include some very small red and white spruce and balsam fir, with alpine shrubs, heaths, azaleas, blueberry, and annual wildflower species (fig. 12.3).

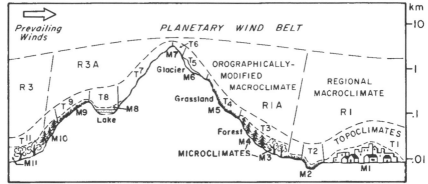

Note: The boundary between the orographically-modified macroclimates RIA and
R3A will vary seasonally and synoptically.

Fig. 12.1. Different microclimates, in general, generated by mountains relative to regional wind flow and regional macroclimate. This variability will occur across most of the alpine areas of New England, with differing levels of variability depending on elevations of the individual peaks and crests of the ranges. Not all mountainous areas will have all of the microclimates depicted in this figure. Modified from Barry (1981, *Mountain Weather and Climate*, Methuen, Routledge; fig. 1.4, p. 12). Used with permission from Taylor & Francis Books Ltd.

Fig. 12.2. Panoramic view of the White Mountains and the Mount Washington Hotel showing the change in vegetation from the larger conifers and hardwoods at the base of the mountains to the treeless tundra at the crest of the mountain range. Lighter spots at the crest of the hills are areas with tundra vegetation. Used with permission from David Metsky as at http://www.cs.dartmouth.edu/whites/photos/mwh1.jpg.

Many mountain summits across New England are of high relief, that is, reach heights well above the valley bottom (table 12.1). Their elevations range up to 6,288 feet at Mount Washington, with eight peaks above 5,000 feet and sixty-seven peaks above 4,000 feet. Mount Washington is clearly the crown jewel of the New England summits, and it is the second-highest peak in the Appalachian Mountain chain, bowing only to Mount Mitchell, North Carolina, at 6,684 feet (fig. 12.4). Although New England summits all share in the basic mountain weather characteristics (lower temperatures with enhanced precipitation and wind), they each display some uniqueness. This uniqueness results from complex interactions between the height, shape, and relative location of the mountain with larger-scale weather systems such as nor'easters, Alberta Clippers, or approaching hurricanes, thereby creating highly varying responses between individual summit locations to even the same weather system. For example, during the heavy snows from 22 to 28 February 1969, Mount Washington recorded a whopping 99 inches of snow, while Pinkham Notch received 77 inches (Kocin and Uccellini, 1990; and fig. 15.9). The next-highest snow total in New England associated with this storm was 56 inches in the mountains of northern Maine. Snowfall totals dropped off quickly from the mountains of New Hampshire and Maine, as the western slopes of the mountains in central Vermont (that is, the lee side relative to this storm system) only received about 10 inches of snow (fig. 15.9).

An understanding of how weather can vary with altitude has taught experienced hikers in New England to be prepared for almost anything weatherwise

Fig. 12.3. Close-up of tundra vegetation (*top*) within the alpine zone of Mount Washington, and view of the overall tundra landscape (*bottom*). The Mount Washington Observatory is on the summit at the left side of the lower picture. Upper photo used with permission from Norbert Woehnl as at http://www.nobby.de/e_unea-s.htm?45; lower photo used with permission from David Metsky as at http://www.cs.dartmouth.edu/whites/photos/lakes7.jpg.

Table 12.1
Mountains over 4,000 feet
in New England

Mountain	Elevation (feet)
NEW HAMPSHIRE	
1 Washington	6,288
2 Adams	5,774
3 Jefferson	5,712
4 Monroe	5,384
5 Madison	5,367
6 Lafayette	5,260
7 Lincoln	5,089
8 South Twin	4,902
9 Carter Dome	4,832
10 Moosilauke	4,802
11 Eisenhower	4,780
12 North Twin	4,761
13 Carrigain	4,700
14 Bond	4,698
15 Middle Carter	4,610
16 West Bond	4,540
17 Garfield	4,500
18 Liberty	4,459
19 South Carter	4,430
20 Wildcat	4,422
21 Hancock	4,420
22 South Kinsman	4,358
23 Field	4,340
24 Osceola	4,340
25 Flume	4,328
26 South Hancock	4,319
27 Pierce (Clinton)	4,310
28 North Kinsman	4,293
29 Willey	4,285
30 Bondcliff	4,265
31 Zealand	4,260
32 North Tripyramid	4,180
33 Cabot	4,170
34 East Osceola	4,156
35 Middle Tripyramid	4,140
36 Cannon	4,100
37 Hale	4,054

Table 12.1 *(continued)*

Mountain	Elevation (feet)
38 Jackson	4,052
39 Tom	4,051
40 Wildcat D.	4,050
41 Moriah	4,049
42 Passaconaway	4,043
43 Owl's Head	4,025
44 Galehead	4,024
45 Whiteface	4,020
46 Waumbek	4,006
47 Isolation	4,004
48 Tecumseh	4,003
MAINE	
1 Katahdin, Baxter Peak	5,268
2 Katahdin, Hamlin Peak	4,756
3 Sugarloaf	4,250
4 Crocker	4,228
5 Old Speck	4,170
6 Bigelow, West Peak	4,145
7 North Brother	4,151
8 Saddleback	4,120
9 Bigelow, Avery Peak	4,090
10 Abraham	4,050
11 South Crocker	4,050
12 Saddleback, the Horn	4,041
13 Redington	4,010
14 Spaulding	4,010
VERMONT	
1 Mansfield	4,393
2 Killington	4,235
3 Camel's Hump	4,083
4 Ellen	4,083
5 Abraham	4,006

From http://www.leisurelybackpacker.com/ ne4000.htm as published by the Appalachian Mountain Club (AMC) in the 26th edition of the *White Mountain Guide.*

Fig. 12.4. View of the Mount Washington massif. Used with permission from David Metsky as at http://www.cs.dartmouth.edu/whites/washington.html.

when in alpine terrain. Even when the forecast in Boston (sea level) and Montpelier, Vermont (515 feet above sea level) calls for fair weather with mild temperatures, mountainous environments can experience thunderstorms or snow, depending on the season. Furthermore, these adverse conditions can occur during all four seasons. Numerous examples have been documented of surprise snow events at altitude, even in summer. For instance, Mount Washington measured over 5 inches of snow in twenty-four hours on 30 June 1988, 1.1 inches on 1 July 1957, and 2.5 inches on 29 to 30 August 1965. Table 12.2 summarizes the climatology of the summit of Mount Washington as recorded at the Mount Washington Observatory. Furthermore, Mount Washington recorded 95.8 inches of snow during the month of May 1997. This amount was nearly twice that of the previous record snowfall total for May. Also note that the highest temperature ever recorded on Mount Washington is only 72°F and that the average afternoon high temperature in July is 54°F. Many tourists headed up the auto road in mid-summer are quite surprised by the cool temperatures on the summit; the Mount Washington gift shop is well stocked with warm clothing.

Even without summer snows, late-season snowfall, along with the cooler temperatures at high elevations, allows for snow conditions to persist well into the late spring and early to possibly mid-summer. Late-season skiing in Tuckerman Ravine, a cirque carved into the east side of Mount Washington, then becomes a tourist destination unto itself (see chapter 4 for an explanation of a cirque). Expert skiers from across the region make the pilgrimage to Pinkham Notch, New Hampshire, to hike to the headwall of Tuckerman and ski

Table 12.2
Climatic normals for the Mount Washington Observatory, 1961–1990,
and record extremes from 1933 to 2001

TEMPERATURE (°F)

	Normal daily maximum	Normal daily minimum	Normal monthly average	Record high (Year)	Record low (Year)
January	12.3	−4.6	3.9	47 (1995)	−47 (1934)
February	13.1	−3.2	5.0	43 (1981, 99)	−46 (1943)
March	20.1	5.4	12.8	54 (1998)	−38 (1950)
April	28.7	16.1	22.4	60 (1976)	−20 (1995)
May	41.0	28.5	34.8	66 (1977)	−2 (1966)
June	49.4	37.9	43.7	71 (1933)	8 (1945)
July	53.6	43.0	48.3	71 (1953)	24 (2001)
August	51.8	41.6	46.8	72 (1975)	20 (1986)
September	45.9	34.6	40.3	69 (1999)	9 (1992)
October	36.4	24.2	30.3	59 (1938)	−5 (1939)
November	27.3	13.9	20.6	52 (1982)	−20 (1958)
December	17.1	0.6	8.9	45 (1992)	−46 (1933)
Annual	33.1	19.8	26.5	72 (Aug 1975)	−47 (Jan 1934)

PRECIPITATION (WATER EQUIVALENT, INCHES)

	Normal	Maximum monthly/annual (Year)	Minimum monthly/annual (Year)	Maximum in 24 hours (Year)
January	7.94	18.23 (1958)	1.29 (1981)	4.85 (1986)
February	8.56	25.56 (1969)	0.98 (1980)	10.30 (1970)
March	8.97	15.98 (1977)	2.15 (1946)	6.45 (1999)
April	8.17	15.21 (1988)	2.19 (1959)	8.30 (1984)
May	7.51	19.00 (1997)	1.78 (1951)	4.60 (1967)
June	7.82	16.00 (1973)	2.43 (1979)	6.50 (1973)
July	7.08	16.58 (1996)	2.69 (1995)	7.37 (1969)
August	8.24	20.69 (1991)	2.46 (1996)	6.63 (1991)
September	7.38	15.47 (1994)	2.74 (1948)	5.38 (1985)
October	7.19	21.25 (1995)	0.75 (1947)	11.07 (1996)
November	10.38	19.56 (1983)	2.31 (1939)	6.07 (1968)
December	9.72	17.95 (1973)	1.49 (1955)	8.64 (1969)
Annual	98.98	130.14 (1969)	71.34 (1979)	11.07 (Oct 1996)

Table 12.2 *(continued)*

SNOW, ICE PELLETS, HAIL (INCHES)

	Record mean	Maximum monthly (Year)	Maximum in 24 hours (Year)
January	40.1	94.6 (1978)	24.0 (1978)
February	40.7	172.8 (1969)	49.3 (1969)
March	42.5	98.0 (1970)	27.4 (1969)
April	30.9	110.9 (1988)	27.2 (1988)
May	10.3	95.8 (1997)	22.2 (1967)
June	1.2	8.1 (1959)	5.1 (1988)
July	Trace	1.1 (1957)	1.1 (1957)
August	0.1	2.5 (1965)	2.5 (1965)
September	1.9	7.8 (1949)	7.7 (1986)
October	11.8	39.8 (2000)	17.0 (1969)
November	40.4	86.6 (1968)	25.0 (1968)
December	42.6	103.7 (1968)	37.5 (1968)
Annual	254.0	566.4 (1968/1969 winter season)	

WIND (MPH)

	Mean speed	Normal direction	Peak gust (Year)	Direction
January	46.3	W	173 (1985)	NW
February	44.5	W	166 (1972)	E
March	41.6	W	180 (1942)	W
April	36.1	W	231 (1934)	SE
May	29.7	W	164 (1945)	W
June	27.7	W	136 (1949)	NW
July	25.3	W	154 (1996)	W
August	25.1	W	142 (1954)	ENE
September	29.1	W	174 (1979)	SE
October	33.8	W	161 (1943)	W
November	39.7	W	163 (1983)	NW
December	44.8	W	178 (1980)	NW
Annual	35.3	W	231 (1934)	SE

From www.mountwashington.org/weather/normals.html after the National Climatic Data Center. Extremes from the time that record collection began at the Mount Washington Observatory in 1933. Used with permission from the Mount Washington Observatory.

down (fig. 4.2). It is not uncommon to see hundreds of cars parked along Route 16, near Wildcat Mountain, left by those who are hiking up to either ski or hike back down Tuckerman. These ventures, however, do not come without their share of dangers, which include snow avalanches and treacherous ski conditions. The Presidential Range has claimed over fifty lives since 1970 (http://www.lexicomm.com/whites/deaths.html, as of the time of this writing).

When the temperature cools, it reduces the ability of air to carry along moisture as a vapor. With cooling, the vapor condenses into tiny liquid droplets, which form clouds. Think about a can of chilled soda in a warm room; the "sweating" is caused by the reduction in temperature of the micro-environment around the can forcing moisture from the air to condense onto the can surface. Air that is lifted to higher elevations is similarly chilled, and condensation commences to form clouds. Water vapor in the atmosphere over the mountains of New England, and particularly over Mount Washington, is often supercooled during winter. However, condensation does not occur until the vapor comes into contact with a solid surface, forming rime. Rime has the appearance to the naked eye of needle-like frost. The buildings of the Mount Washington observatory act as nuclei for the formation of rime (fig. 12.5). Given the prevalent winds on the summit of Mount Washington, rime will build up in a "flag-like" formation pointing into the wind as layer upon layer of rime forms (fig. 12.6).

High elevations are also noted for strong and persistent winds, with Mount Washington serving as the national, or even international, focal point for discussions on windiness. This distinction results from Mount Washington holding the world record for the fastest surface wind speed ever measured (231 miles per hour—before the anemometer broke). Although it is occasionally calm on Mount Washington, winds average about 25 miles per hour in summer and about 45 miles per hour in winter (table 12.2). However, it is noteworthy that every month has experienced winds over 100 miles per hour and winds of this velocity are quite common, especially in winter.

There are several reasons for the high winds on Mount Washington. First, it lies near or within most of the prominent storm tracks in North America, including those for Alberta Clippers and nor'easters. Second, the Presidential Range within the White Mountains runs roughly north-south, which reduces the vertical depth of the atmosphere (troposphere), thereby forcing westerly airflow through a narrower passage over the mountain. Third, Mount Washington rises several hundred feet over the next-highest summit. Hence, Mount Washington is like an island poking high into the upper-level wind systems like the jet stream, where friction from Earth's surface is not a factor in slowing down the wind. There are also no peaks of near-equivalent altitude to the west to help break up the wind until one reaches the high plains on the eastern slope of the Rocky Mountains.

Fig. 12.5. Rime formations on various parts of the Mount Washington Observatory, January 1997. Photos by Greg Zielinski.

Although we suspect that many tornadoes have wind speeds greater than the record on Mount Washington, those winds are yet to be measured in a comparable way. Therefore, Mount Washington holds the claim of the highest wind speed ever measured, though this is not necessarily the highest wind ever on the face of Earth. The world record wind of 231 miles per hour is still a most impressive feat for the mountain and helps in making the claim that the mountain is home to "the world's worst weather."

Fig. 12.6. Sampling rime on the deck of the Mount Washington Observatory in sustained 100-plus mile per hour winds and a temperature of –30°F, January 1997. The peak gust recorded on this day was 138 miles per hour. Photo by Greg Zielinski.

The record wind was recorded shortly after noon on 12 April 1934. Actually, two readings of 231 miles per hour were measured within a fairly short period of time by Alexander McKenzie (McKenzie, 1984). The weather setting for the day included a frontal system located near the mid-Atlantic states and Chesapeake Bay. Therefore, there was cooler air at the surface to the north of the front (including New England), with warmer and moister air to the south (Dave Thurlow, personal communication). It is believed that the warmer southerly air glided up and over the cooler surface air over New England. Such a situation, which is not all that uncommon, creates a boundary layer over the cold surface air and allows the warm air to move over a surface that is very smooth, hence winds can move quite rapidly. The air that arrived on Mount Washington to set the record arrived from the southeast and was also funneled and constricted by the southeast-northwest oriented valley on the southeastern side of the mountain. Clearly, this further increased the winds as they reached the summit on 12 April 1934. What is most curious about this event is that southeast winds are the least common on the summit, as the large majority of the wind comes from westerly directions.

While we have focused on Mount Washington and its alpine environment, other mountains in the region share many of the same characteristics we have mentioned. Dangers lurk here, but the beauty and tranquility of the mountains of New England continue to attract tourists to their summits. While a good adventure can make you feel alive, make certain to use some wisdom and do not throw caution to the wind as you venture into the mountains of New England.

Summary

The central mountains of New England comprise a significant portion of the landscape and as a result, they play a major role in forcing the region's climate.

Foremost, temperatures cool with an increase in elevation. Consequently, distinct changes in vegetation come with elevation. Many of the higher mountains in the region, such as Mount Washington, are dominated by tundra vegetation on the upper slopes. The highest temperature ever recorded on the summit of Mount Washington is only 72°F. Wind speeds also increase with elevation, given less friction with Earth's suface as one goes up in height. This aspect contributed to the record 231 miles per hour wind recorded at the summit of Mount Washington in 1934. Alpine regions play an important role in the distribution of precipitation, as windward slopes will see enhanced precipitation levels with an individual storm compared to the lee side and often compared to adjacent valleys. Mount Washington receives, on average, 99 inches of precipitation per year and 254 inches of snow per year.

Part V

The Weather Events that Influence the Lives of New Englanders

Everyone talks about the weather, but nobody does anything about it.
—MARK TWAIN

When Mark Twain made the previous statement (see p. 139), we cannot help but wonder if he was referring to specific weather events, because as the climate of a region changes more slowly, it generally is easier for the overall population to adjust to that change. In general, large-scale climatic changes should be slow enough that society as a whole can adapt accordingly. However, certain segments of society may not be able to adjust as easily as others following specific meteorological or climatological events. For example, many farmers cannot cope with two to three years of bad crops following adverse climatic conditions, such as two successive very dry summers. Smaller ski resorts might file for bankruptcy following several years of low snowfall totals.

More importantly, an infrequent but high magnitude single weather event has the potential to completely alter one's life in a matter of several hours or days (e.g., Watson, 1990). It is these extreme events that probably have the greatest impact on humans and that leave long-lasting impressions. These events also can alter the landscape to the point that changes may be observed through one's lifetime (i.e., on the order of several decades).

Bringing up a particular event for discussion can trigger individuals' memories of that particular event. For example, in 1999, the Boston Museum of Science presented a theme weekend entitled "Wild Weather Weekend." We presented information on past blizzards in New England and conducted an informal survey of visitors' most memorable snowstorm. Given that this survey was done in Boston, the 1978 Blizzard was, by far, the snowstorm people remembered most. Almost every individual had a story to tell about that blizzard. One person remembered taking an hour to go down the block to check on relatives; another had to climb out of a second story window to get out of the house. Stories like these undoubtedly come with every large weather event in New England's past, as these extreme phenomena are the type of weather events that stick with you. After all, the most common topic of conversation at Tut's General Store and Restaurant, Waterford, Maine, after discussions about community events, is the weather (Yankee Magazine, June 1999, p. 41). Entire books have been dedicated to New England weather events and their impact on New Englanders (see Thomas, 1990).

Certainly, New England has had its share of famous weather events, primarily in the form of large storms. In fact, one of the top ten storms of the twentieth century as chosen by experts from the Weather Channel—the 1938 Hurricane—primarily had a major impact on New England, while another storm had a major impact on New England together with the rest of the eastern *seaboard—the March 1993 Superstorm. The 1938 Hurricane was voted the fifth-greatest storm of the millennium, while the March 1993 Superstorm was voted the third-greatest storm. The Blizzard of '78 was given honorable mention as was the 1991 All Hallows Eve storm ("The Perfect Storm"). A similar survey by* Weatherwise Magazine *placed the 1993 Superstorm as the fourth-greatest weather and climate event of the last millennium, while the 1938 Hurricane was tenth out of ten. The Blizzard of '78 and the 1998 ice storm were given honorable mention. Surely, these findings indicate that New England has its share of historical weather events. Our top ten weather events of this past century are listed in table V.1. We discuss many of these in more detail in this part of the book, as well as hallmark events from other centuries.*

We describe six types of weather events or related groups of events in this part of the book. Each chapter defines and describes the origin of that partic-

Table V.1

Top ten weather events of the twentieth century in New England

Date	Type of event	Major impact
21 Sept. 1938	hurricane	17-foot storm surge on Rhode Island and Connecticut coast; high wind and rain inland; 600+ deaths
5–7 Feb. 1978	nor'easter/blizzard	high snow and winds southeast New England with up to 50 inches in Rhode Island
17–19 Aug. 1955	hurricane	24-hour record rainfall of 18.15 inches in Westfield, Massachusetts
mid-Mar. 1936	rain on snow/flood	"All New England Flood" produced by heavy rain fall on a deep snowpack
5–9 Jan. 1998	ice storm	widespread tree damage, downed power lines leading to extensive losses
9 June 1953	tornado	F4 tornado in Worcester produced 200–260 mph winds and 90 deaths
12 Apr. 1934	wind	231 mph wind at Mount Washington Observatory is highest wind ever recorded
20–21 Oct. 1996	nor'easter/rain event	all-time record rainfall for New Hampshire and Maine with 19.2 inches at Camp Ellis, Maine
22–28 Feb. 1969	nor'easter/bizzard	99 inches of snow on Mount Washington and 77 inches at Pinkham Notch
3–4 Nov. 1927	rain event	nearly 10 inches of rain across much of Vermont leading to 84 deaths

ular type of event, the potential impacts of those events on New Englanders, record setting events, and specific details about the hallmark events of each type, and possibly other memorable events of the particular weather type. For example, a particular storm or event does not necessarily have to be a record setter to be memorable! We also point out that the events we discuss either occurred over the last one hundred years, when most instrumental records began to be kept, or occurred over roughly the last two to four hundred years, when historical accounts are available following European occupation of New England. Even greater events probably swept New England prior to European settlement, but we have less information about them.

Chapter 13

Heat Waves and Cold Spells

Thirty below this morning. Seems like it might get cold . . . or . . . Ninety already and the sun's not over the mountain.   —DONALD HALL SPEAKING ABOUT
QUOTES BY HIS GRANDMOTHER

The first type of events we focus on are those related to extremes in temperature, as so succintly stated in the above quote. Although the climate system is made up of many components, temperature probably is the one parameter that many individuals think about the most. However, when it comes to specific events, temperature extremes probably do not get as much recognition as other types of events, perhaps, in part, because the conditions responsible for temperature extremes occur over a period of days. Furthermore, the impact of temperature extremes in New England may not be as "newsworthy" as in other parts of the country. This would be especially true for areas like the central United States, where extensive heat waves can cause a high number of deaths and damage to livestock and other aspects of the agricultural industry. Another reason why heat waves and cold spells may not get as much recognition in New England is that, although they do exist, they have a tendency to be more short-lived here than in other parts of the country. Thus, the socioeconomic impact is not as widespread or as great as in other parts of the country. Nevertheless, heat waves and cold spells can be dramatic in New England. We discuss these two end members of temperature-related events in this chapter.

Heat Waves

A. T. Burrows in 1900 defined a "hot wave" as three successive days with temperature equal to or exceeding 90°F. This is a rather arbitrary definition, as temperatures of 88 or 89°F would feel about the same to the human body. Moreover, when temperatures of this magnitude occur in New England, they are almost always accompanied by high humidity. Heat index values—the combination of temperature and humidity—may easily exceed 100°F on those

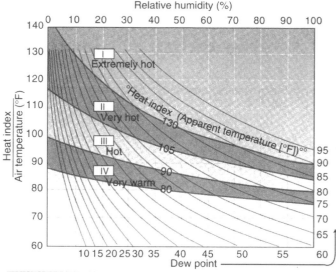

Category	Apparent temperature (°F)	Heat symptoms
I	130° or higher	Heatstroke/ Sunstroke highly likely with continued exposure
II	105° - 130°	Sunstroke, heat cramps, or heat exhaustion likely. Heatstroke possible with prolonged exposure and physical activity.
III	90° - 105°	Sunstroke, heat cramps, and heat exhaustion possible with prolonged exposure and physical activity.
IV	80° - 90°	Fatigue possible with prolonged exposure and physical activity.

Fig. 13.1. Chart showing the relationship between temperature, dew point, and relative humidity with the resulting heat index. The lower part of the chart shows the potential impacts of each heat index category (I to IV) identified in the upper chart. The heat index can be determined by knowing the air temperature and either the relative humidity or the dew point. To determine the heat index, find the point where the actual air temperature (left axis) intersects the relative humidity value (upper axis) or where the air temperature intersects the curved lines that represent dew point (lower and right axis). The intersection of either of these lines will fall in one of the heat index zones in the center of the figure. From Danielson et al. (1998; fig. I.1, p. 431). Used with permission from The McGraw-Hill Companies.

days even with ambient air temperatures in the low 90s. Heat index is determined by combining air temperature with the amount of moisture that is present in the air (fig. 13.1). One measurement of the amount of moisture is the dew point. Anytime the dew point exceeds 57°F, temperatures feel hotter than the thermometer registers because the moisture in the air prevents sweat from evaporating from your skin surface. Dew points in the mid-60s and above lead to those oppressive days when one can be sweating with very little activity. Dew points in the 70s are tropical. Only rarely do temperatures reach the 90s in New England without high humidity. In cases when we do get high temperatures and dry air, a strong high pressure system is located southwest of New England. Air flowing around the west side of the high may originate in the central Great Plains and partly in the Southwest (fig. 2.11). Consequently, New England will be under the influence of hot continental air. Most of the time, heat waves in New England are associated with a strong Bermuda-Azores High and a high pressure system more south-southwest to southeast of New England. This scenario allows maritime tropical air to penetrate all the way into New England from the Gulf of Mexico or the subtropical portion of the Atlantic Ocean (fig. 2.7). Flow from these areas brings in high humidity and often can persist for up to a week. The stronger the high pressure system, the harder it is to "move it away" from the eastern United States. This anticyclone thus becomes a blocking high, stopping all weather sytems from moving into the eastern United States. In fact, it may force storms to move well to the north of the center of the high. In particular, when upper air patterns, or westerlies, are fairly weak or disrupted, these high pressure systems can become very well entrenched over the southeastern United States (fig. 2.9). Such a situation can lead to an extended heat wave.

Areas adjacent to the coast are less susceptible to heat waves because of the formation of sea breezes, particularly during the afternoon. We described how a sea breeze forms in chapters 2 and 5, although its formation is not consistent over the entire New England coastline on any single summer day. A major consideration is how strong the regional winds are and in what direction are they blowing. If a strong southwesterly flow exists, then a sea breeze is especially likely to form along the southern New England coast (Connecticut, Rhode Island, Cape Cod), along the southern edge of the capes (e.g., Cape Ann, Massachusetts), and Down East Maine. However, if winds are strong out of the west—mostly off the continent—sea breezes may be suppressed throughout New England. In cases where there are no strong regional surface winds, sea breezes are most likely to develop across much of the New England coastline, because of differential heating between the land and the ocean. In this scenario, local winds are much more dominant than the regional flow.

The record high temperature measured in New England was 107°F in New Bedford and Chester, Massachusetts, on 2 August 1975, that is, "Hot Saturday."

Record highs for individual states are shown in table 9.1. Rhode Island also recorded its hottest day ever on "Hot Saturday." Every state had readings over 100°F on this day with temperatures such as 104°F in Jonesboro, Maine, 103°F in Hanover, New Hampshire, and Portland, Maine, 102°F in Boston, Massachusetts, and 101°F in Hartford, Connecticut, Concord, New Hampshire, and Bar Harbor, Maine. The 100-plus temperatures in Down East Maine certainly indicate no southerly sea breeze blew on that day. It did not matter whether one was inland or at the beach, "Hot Saturday" was hot throughout the region. Of the many heat waves that have hit New England, one of the more impressive was the one that struck in July of 1911. Successive afternoon high temperatures from 1 July to 12 July at Keene, New Hampshire, were as follows: 91, 95, 104, 103, 103, 101, 88, 91, 99, 102, 99, 95°F. Considering that three successive days constitute a heat wave, this 12-day event is in no uncertain terms a "mega-heat wave." This event also produced the record high temperature ever measured in New Hampshire: 106°F in Nashua on 4 July 1911, as well as the record high of 105°F in Vernon, Vermont (table 9.1). Other notable heat waves have occurred over the past three to four hundred years. One such event occurred in late June and early July of 1825. Rev. Ezra Stiles of Yale College in New Haven, Connecticut, recorded four successive days over 90°F from 29 June to 2 July, and another four days from 8 July to 11 July. Temperatures were above 87°F for nine successive days from 26 June to 4 July in that same year (Ludlum, 1976). Ludlum documents other heat waves in June 1749, July 1825, June 1925, and August 1948.

The Heat-Wave Killer

Once a strong Bermuda High becomes entrenched along the East Coast, it often takes an equally strong high pressure system in Canada to push a cold front through New England to break the heat wave. However, sometimes a high pressure system will drift northeast of New England. This high pushes cool maritime air off the Atlantic Ocean into New England (fig. 13.2). This push of cool air is strong enough that a distinct cold frontal boundary will form, separating the cool, moist air originating to the northeast of New England from the hot, moist air originating south of New England (figure 13.3). The interesting aspect of this type of cold front is that it moves primarily from northeast to southwest across the region. Because this movement is contrary to the west-to-east movement of most fronts, it is referred to as a backdoor cold front. Backdoor cold fronts often occur in the spring, when the cool air may push up against the Appalachian Mountains well into the Middle Atlantic states. However, late in the summer, this cool air often will encroach only into New England. The hot, humid air may eventually "win out" and displace the cooler air back to the northeast or east (as a warm front), but one or two days of tem-

Fig. 13.2. Classic back-door cold front bringing maritime polar air (mP) into the United States, east of the Appalachian Mountains. The mountains essentially dam this cool air to the east. The southwestern extent of the backdoor cold front, as shown here, often occurs during mud season. However, it is usually only the New England area that benefits from this scenario during the summer (see text and fig. 13.3). From Danielson et al. (1998; fig. 9.11, p. 247). Used with permission from The McGraw-Hill Companies.

peratures in the 70s or low 80s, accompanied by lower dew points (i.e., less-humid air), are most comforting to New Englanders.

A classic example of this mid-summer backdoor front occurred during August 2001 (fig. 13.3). On 6 and 7 August, a high pressure to the southwest of New England began to pump in hot and humid air. High temperatures over a good portion of New England were in the 90s by 7 August. A high pressure system moving north of New England eventually made it to a point northeast of the region by the eighth. At this time, cooler air began to push into New England as a backdoor cold front eventually moved through most of the area (fig. 13.3). Within a day, however, the front moved back to the north as a warm front, again bringing the hot and humid air into all of the region. A similar scenario occurred in July 1997. Much of the East Coast, including New England, was sweltering under an oppressive heat wave. Temperatures exceeded 90°F for two to three days from 13 July to early on the fifteenth, when a weak back door cold front moved through Maine and New Hampshire. Its progress stopped around central Massachusetts before it eventually returned to the northeast as a warm front. The result was a drop in maximum temperatures in New Hampshire and Maine from the 90s on 15 July to the high 70s on 16 July. Temperatures eventually made it back into the 90s on 18 July in New Hampshire and Maine. However, areas in western Massachusetts, Connecticut, and Rhode Island remained to the southwest of this cool air, producing five successive days in the 90s—a true heat wave. Northeastern New England, on the other hand, failed to have three successive days in the 90s; an official heat wave did not occur in New Hampshire and Maine early that July. The backdoor front can truly be a "heat-wave killer" in New England.

Cold Spells

No definition for a true cold spell has been designated like the one that exists for heat waves. The coldest of cold spells occur when high pressure in Canada

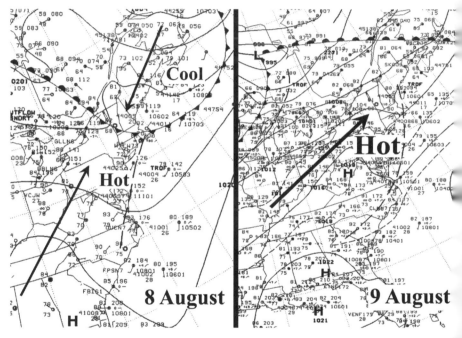

Fig. 13.3. Example of a backdoor cold front that advances through only part of New England during t summer. On the left-hand map, the backdoor cold front has migrated through Maine, New Hampshire, most of Vermont, and about half of Massachusetts, reaching its furthest extent as shown here. Neither Connecticut or Rhode Island felt the flow of cooler air off the Atlantic Ocean, as they remained in the h and humid air from the southwest (shown by lower left arrow). The northeast air flow (shown by upper right arrow) originated from a high pressure system positioned northeast of New England, just off the upper part of the picture. This map is a snapshot for 1000 hours, 8 August 2001. In the right-hand map, the backdoor front has retreated northeast as a warm front, having moved completely through New Eng land by 1000 hours, 9 August 2001. The entire region is now under the influence of hot and humid air originating from around the high pressure systems located in the southern United States. Maps from National Climatic Data Center (National Oceanic and Atmospheric Administration/Department of Commerce) web s (http://www4.ncdc.noaa.gov/cgi-win/wwcgi.dll?wwAW~MP), at the time this was written. Maps originate from the National Center for Environmental Prediction (NCEP).

Temperature (°F)																		
Calm	40	35	30	25	20	15	10	5	0	-5	-10	-15	-20	-25	-30	-35	-40	-45
5	36	31	25	19	13	7	1	-5	-11	-16	-22	-28	-34	-40	-46	-52	-57	-63
10	34	27	21	15	9	3	-4	-10	-16	-22	-28	-35	-41	-47	-53	-59	-66	-72
15	32	25	19	13	6	0	-7	-13	-19	-26	-32	-39	-45	-51	-58	-64	-71	-77
20	30	24	17	11	4	-2	-9	-15	-22	-29	-35	-42	-48	-55	-61	-68	-74	-81
25	29	23	16	9	3	-4	-11	-17	-24	-31	-37	-44	-51	-58	-64	-71	-78	-84
30	28	22	15	8	1	-5	-12	-19	-26	-33	-39	-46	-53	-60	-67	-73	-80	-87
35	28	21	14	7	0	-7	-14	-21	-27	-34	-41	-48	-55	-62	-69	-76	-82	-89
40	27	20	13	6	-1	-8	-15	-22	-29	-36	-43	-50	-57	-64	-71	-78	-84	-91
45	26	19	12	5	-2	-9	-16	-23	-30	-37	-44	-51	-58	-65	-72	-79	-86	-93
50	26	19	12	4	-3	-10	-17	-24	-31	-38	-45	-52	-60	-67	-74	-81	-88	-95
55	25	18	11	4	-3	-11	-18	-25	-32	-39	-46	-54	-61	-68	-75	-82	-89	-97
60	25	17	10	3	-4	-11	-19	-26	-33	-40	-48	-55	-62	-69	-76	-84	-91	-98

Wind (mph)

Frostbite Times ▦ 30 minutes ▦ 10 minutes ▢ 5 minutes

$$\text{Wind Chill (°F)} = 35.74 + 0.6215T - 35.75(V^{0.16}) + 0.4275T(V^{0.16})$$

Where, T= Air Temperature (°F) V= Wind Speed (mph)

Effective 11/01/01

Fig. 13.4. New wind-chill index developed by the National Weather Service in 2001. The chart also shows the amount of time in which frostbite can occur on exposed skin (shaded values on right-hand side) and the formula used to derive the wind chill index. Values in this chart are lower than in previous versions. Chart may be found at http://205.156.54.206/om/windchill/ (National Oceanic and Atmospheric Administration/Department of Commerce).

pushes air from over the north pole into New England (fig. 2.10). This "Siberian Express" will drop nighttime temperatures well below 0°F over a period of several nights (see chapter 7 for details). In fact, daytime temperatures sometimes do not reach 0°F or they may not get out of the single digits. During the initial movement of this frigid air into New England, brisk northwest winds will drop wind chill readings to dangerous levels (fig. 13.4). Wind chill is defined as the temperature sensation on exposed skin adjusted to conditions as if there was no wind. For instance, 30 to 40 mile per hour wind gusts will make the 0°F readings feel like −40 to −50°F on exposed skin, hence −40°F is the wind chill temperature. Such readings are not uncommon when arctic air is brought into New England during winter. However, minimum temperatures during a cold spell are reached as the high pressure system migrates directly over New England. Snow cover, radiational cooling, and the lack of wind to stir up the atmosphere under this scenario can easily drop nighttime temperatures into the −20s to −30s range or lower. The record low temperature for New England is −50°F in Bloomfield, Vermont. Other record lows for individual states are shown in table 7.2. Areas that usually will experience the coldest temperatures are the valley bottoms, as dense cold air sinks to these low spots at night. Temperatures will then be much lower than the surrounding hill slopes or upland areas.

Although cold spells are not as easily defined as heat waves, frigid temperatures persisted for long periods of time during several winters over the past few hundred years in New England. One such case was the winter of 1917/1918. Nighttime temperatures around New England were consistently below 0°F for the twelve-night period from 26 December to 12 January. Temperatures bottomed out in the −40s several times over that stretch at several sites in northern New England, including Saint Johnsbury, Vermont, and Berlin, New Hampshire (Ludlum, 1976). Temperatures failed to climb above freezing on any of those days. A similar situation occurred in December 1969 through much of January 1970. During that period, the temperature never went above freezing in Burlington, Vermont, with below zero readings every night except two. February 1943 is another example of a frigid month, as places as far south as Nantucket and New Haven fell below zero by the night of 16 February. Although we have not had as many cold winters in recent time, there still have been some record-setting periods (see discussion in chapter 19). As recently as March 2001, record lows were set in several places in Maine. The winter of 2000/2001 was snowy and cold around New England, which led to a very deep and persistent snowpack. This snowpack enhanced nighttime cooling, as the primary reason for the cold winter was daily minima. Maxima for much of that winter were close to average, but average nighttime lows for several months were consistently as much as several degrees below average.

When we think of heat waves and cold spells, we naturally tend to think of summer and winter extremes. On the other hand, sometimes New England experiences extreme warmth during winter or extreme cold during summer. These situations may set records opposite to what would be considered normal during the respective seasons. We introduced two specific examples in chapter 3 when we discussed factors that influence climate on a year-to-year basis. The winter of 1997/1998 will remain in some people's minds because of its record-setting warmth associated with the El Niño event that year, not because it was a cold winter. For instance, temperatures in Portsmouth, New Hampshire, soared to 90°F on 31 March 1998, the warmest day of the entire year.

The opposite scenario holds for the the summer of 1816. That summer is deeply entrenched not only in New England history, but in global history as the "Year without a Summer" following the eruption of Tambora in Indonesia in 1815. Snow in June and frosts in July and August occurred in New England that summer.

The transition seasons also can be marked by early cold spells in fall or early warmth in spring. Extremes during the transition seasons do not seem to make the news or stick in the memory as often as those during the summer or winter. Being transition seasons, the average temperatures for spring and fall months are in the middle of the annual range, thus the extremes do not stand out on an annual basis as much as extremes during winter or summer. Nevertheless, cold outbreaks early in fall or late in spring and the resulting early or late frosts have been known to produce dire consequences for several aspects of New England life. Individuals in agriculture are especially affected by these types of temperature events. For instance, the combined average temperature of autumn 2001 (September to November) and winter 2001 to 2002 (December to February) is the warmest for this six-month period in the 107 years of record for each New England state.

Summary

The dominance of two different high pressure systems leads to heat waves and cold spells in New England. A very strong Bermuda High will block other systems from moving into the eastern United States, thereby continuing to pump hot and humid air into New England. Such a situation led to "Hot Saturday" on 2 August 1975, when temperatures above 100°F were recorded in every state. The twelve-day July 1911 heat wave was a prolonged period of temperatues in the 90s over much of the region. Luckily, a back door cold front often will move through much of New England for just a few days during the summer, acting as a "heat-wave killer." A strong Canadian High that may bring air from directly over the North Pole into New England is capable of lowering temperatures to the –40s. Such cold conditions are enhanced by cloudless, calm nights with a snowpack.

Chapter 14

Droughts and Rainstorms

You fix up for the drought; you leave your umbrella in the house and sally out,
and two to one you get drowned. —MARK TWAIN

Heavy rainfall and drought are on the opposite ends of the moisture continuum. Given New England's predisposition for extreme weather, it has more than its fair share of parching droughts and periods with torrential deluges. Droughts can decimate water supplies and agricultural yields, while intense rainstorms can cause urban flash floods that can paralyze cities and towns with flooded and washed-out roads. Larger storm systems may produce wider-scale river-basin flooding. In either of these cases, monetary losses can be great and lives can be lost, particularly from floods.

Droughts

Droughts result from a lack of precipitation for some prolonged period of time, often because a strong high pressure system prevents storms from entering New England. The most common scenario is when the Bermuda High becomes entrenched over the eastern United States, thus pushing storms to the north of New England—called a blocking high. Certain jet stream patterns also may carry storms to either the north or south of the region. After some time with little or no delivery of moisture, regardless of the cause, the natural system becomes disrupted. Although New England receives around 38 to 48-plus inches of precipitation per year on average, nevertheless it is susceptible to drought conditions (figs. 11.2 and 11.3).

Part of the reason for New England's susceptibility to drought is the nearly equal distribution of precipitation throughout the year (see chapter 11 and fig. 11.5). Thus, drought conditions can occur in any season, although summer drought presents a much greater problem than winter drought. Warm-season droughts are much more of a concern because the higher temperatures in summer enhance evaporation and transpiration by plants, so that the lack of precipitation is compounded by the removal of moisture from the soil and from lakes and ponds to the atmosphere. When a drought begins, moisture stored in the pore space of soil (known as soil water) is first depleted through the process of evapotranspiration. Evapotranspiration is the total loss of water vapor to the air by the combined processes of evaporation and plant transpiration. Initially, vegetation wilts, and it eventually will die if no water is provided. As a drought continues, lake levels, river volumes, reservoirs, and groundwater are all adversely affected. Water shortages then become problematic not only for humans, but for all aspects of the many ecosystems within the region. Furthermore, pollution levels in water tend to increase because the lower water volumes in freshwater systems concentrate the pollutants. In addition, since

Fig. 14.1. April 1998 brush fire in North Berwick, Maine (*top*), one of the potential problems resulting from drought conditions in New England. The bottom photo shows a forest fire warning sign from the Maine Forest Commissioner, 1917. Top photo used with permission from Dave Hilton as at http://www.geocities.com/rainforest/4192/nberwoods.jpg, at the time this was written. Bottom photo used with permission from Dave Hilton as at http://www.geocities.com/rainforest/4192/colby1.jpg.

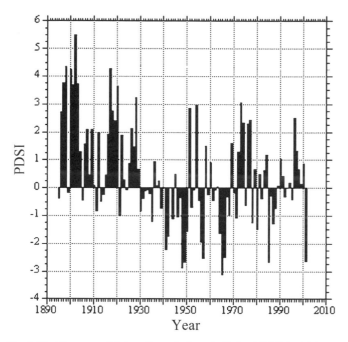

Year

Fig. 14.2. Mean annual Palmer Drought Severity Index (PDSI) for Maine from 1895 to 2001. Positive values reflect years with adequate moisture availability, while negative values reflect years with a water deficit. The lowest annual PDSI on record for Maine is 1965. Interestingly, Maine was probably the least dry of the New England states during the 1965 to 1966 drought. Note the overall trend of wet conditions from 1895 to the late 1920s, dry conditions from the late 1920s to the early 1970s, and the generally wet conditions from the 1970s until 2000. The late 1980s are the major dry period of the last thirty years, although 2001 is now the year with the lowest annual precipitation.

evaporation and transpiration serve to cool the atmosphere, temperatures tend to become even hotter when little moisture remains on the surface: Incoming solar radiation goes toward heating Earth's surface rather than evaporating water. Surface heating is further increased by the lack of afternoon cloud cover, which can lessen extreme temperatures. Another hazard associated with drought is brush and forest fires (fig. 14.1). When trees and other vegetation are under water-shortage stress, they become more flammable and fires can easily rage out of control.

Dry summer conditions were prevalent in New England from about 1930 through 1966, with the worst drought conditions occurring in the early to mid-1960s, with peak drought conditions in 1965 (Leathers et al., 2000). We show a record of drought conditions in New England since 1895 using Maine as the example in figure 14.2. Dry conditions are often quantified using the Palmer

Drought Severity Index (PDSI). The PDSI not only takes into account precipitation, but it considers other factors, such as temperature and evaporation. In this index, values from −1.99 to +1.99 indicate mid-range conditions, −2.0 to −2.99 indicate moderate drought, −3.00 to −3.99 indicate severe drought, and values equal to or less than −4.00 represent extreme drought. Equivalent designations exist for the moist side, that is, +2.00 to +2.99 indicates moderately moist conditions, and so on. The driest year on record, 1965, averaged out with severe drought conditions, and 1966 was designated with moderate drought conditions by the Palmer Index (fig. 14.2). It is noteworthy that of all the New England states, Maine may have fared the best in 1965, yet it was the second worst year for total precipitation across the state. Interestingly, the late 1940s are the second-worst time period under the PDSI designation, almost averaging out under severe drought conditions. In the year 1947, large fires consumed significant parts of Mount Desert Island (Acadia National Park). Other very dry years occurred in the early 1940s, mid-1950s, 1985, and most recently in 2001, the driest year on record for Maine over the last 107 years (fig. 14.3). The third driest year on record for New Hampshire and the fifth driest for Vermont was also 2001.

Although summer droughts ultimately may have more of an impact on New England, winter droughts are very much a problem as well. A lack of snow in the New England mountains affects local ski conditions and a lack of snow

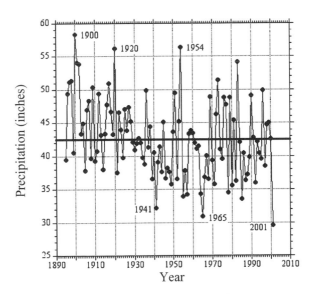

Fig. 14.3. Maine annual precipitation in inches for 1895 to 2001, showing the three driest and wettest years on record. The solid line indicates the 107-year annual average of 42.6 inches.

in Boston tends to keep people from thinking of skiing. As a result, there are many years when snow in the mountains is deep and the skiing great, but with no snow in Boston, the slopes are empty. The opposite can also be true, when there is snow in southern New England and everyone is thinking about skiing, yet the northern mountains are bare. Winter snowcover also provides spring snowmelt, which brings rivers and streams back to life as well as assisting in the recharge of groundwater. The lack of spring recharge to water systems following a winter drought may exacerbate summer drought conditions, leading to multiple-year drought concerns.

Extreme Rainfall

Heavy rainfall is usually made up of an excessive amount of water delivered to the landscape in a very short period of time, that is, a few hours to possibly a few days or more. This rainfall will typically saturate the surface soil layer with water, recharge groundwater aquifers, and probably run off of the surface into rivers, streams, and lakes. How much goes into each of these areas depends on several factors, such as the nature of the soil and how much previous rainfall there has been over the last few days or even the last few weeks. These factors determine how saturated the soil will be at the time of a major rain event and how much runoff will occur.

New England can experience heavy rainfall from three sources: localized storms that are generated by free convection (such as thunderstorms), particularly in summer; synoptic scale mid-latitude cyclones, for example, nor'easters or Alberta Clippers, and associated frontal systems; and hurricanes or tropical storms and their remnants. We briefly discuss the mechanisms that produce each of these types of precipitation.

Afternoon thunderstorms primarily occur in summer, especially when dew points are very high, so that the air is very unstable. They occur under free-convection conditions, when the surface is heated by the sun to the point where a parcel of air floats upward. This updraft forms a cloud that can occasionally develop into a thunderhead (fig. 14.4). Afternoon thunderstorms can produce very heavy rainfall, but the rain typically lasts only minutes to hours and is limited in area. In fact, even when there is heavy rain associated with thunderstorms, these storms are not effective for breaking a drought since the storms often are isolated rather than very widespread. As a result of the short duration, the amount of rain from a thunderstorm is generally less than that produced by larger-scale weather systems. However, an intense thunderstorm that occurs during rush-hour traffic can produce many dangers, including reduced visibility and high water in low places like underpasses. The eastern shore of New England, including the coasts of Maine, New Hampshire, and eastern Massa-

Fig. 14.4. Developing thunderstorm over Bangor, Maine, showing evidence of updrafts along the edge of the storm as indicated by cirrus clouds on right side of picture. A rainbow is evident across the left side of the view. Photo by Greg Zielinski.

chusetts, however, do not experience many summer thunderstorms. In fact, these areas average fewer than twenty days per year with a thunderstorm— fewer thunderstorms than anywhere else in the United States east of the Rocky Mountains (see chapter 9). The reduction in thunderstorms results from a cooling sea breeze (from off of the cold Labrador current) in the afternoon that does not allow the surface in the coastal zone to become warm enough to make a parcel of air rise with any consistency.

Mid-latitude cyclones are areas of low pressure that derive their energy from highly contrasting air masses. Components of a typical cyclonic system in New England would include warm moist air from the Gulf of Mexico or tropical Atlantic Ocean that is in contact with cold and dry air from northern Canada. These opposing types of air are divided along frontal boundaries, where precipitation can be produced (fig. 14.5). Cold fronts tend to have more intense rainfall, often including thunderstorms over a narrow band ahead of the approaching front. Warm fronts usually produce more gentle rains, but over a

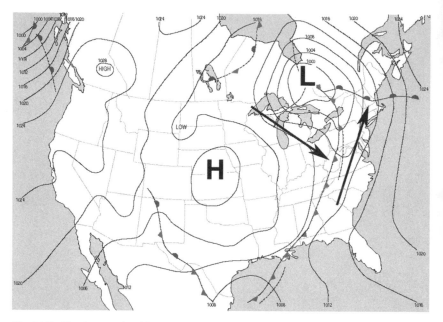

Fig. 14.5. Classic example of frontal systems approaching New England associated with low pressure system to the west of the region (see fig. 2.13). The warm front will move through New England initially, bringing in warm, humid air from the south, followed by the cold front bringing down cool, dry air from Canada.

wider area. These storm systems often produce the comma-shaped cloud pattern seen in satellite imagery (fig. 14.6).

Although New England does not experience tropical weather very frequently, it does occur, and the impacts are often catastrophic. These systems typically originate over the Atlantic Ocean and then move up the warm Gulf Stream ocean current. Depending on atmospheric pressure patterns, the storm can either move out to sea, hit the southeast to mid-Atlantic coast of the United States, or go due north and strike New England (fig. 14.7). These storms are huge storage tanks of moisture, even after winds speeds have subsided. Rainfall totals associated with these tropical systems can easily reach 5 inches in less than a day. Chapter 18 provides more details about hurricanes.

Table 14.1 displays the largest precipitation events recorded in New England over the past sixty years. The largest single-day precipitation event recorded in New England was 18.15 inches at Westfield, Massachusetts, which was produced by Hurricane Diane in late August of 1955. In all, this single event produced 19.75 inches of rainfall at Westfield over three days (18 to 20 August 1955), the single largest rainfall event in New England since instrumental records began in the 1920s. One-day rainfall totals from this event were in excess

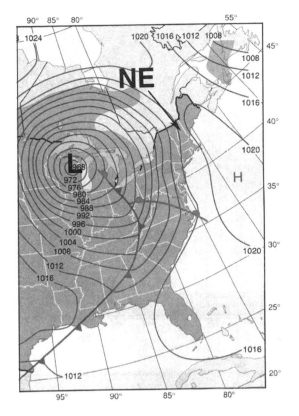

Fig. 14.6. Strong storm system and associated fronts (*top*) approaching New England with the typical comma cloud pattern (*bottom*) associated with these mid-latitude storm systems. The center of low pressure is marked by L on each figure, and the location of New England is shown by NE. This was the storm system position as of 3 April 1982. Modified from METEOROLOGY: 5/E, THE ATMOSPHERE AND THE SCIENCE OF WEATHER by Moran/Morgan, © 1997 (fig. 11.10, p. 265). Reprinted by permission of Pearson Education, Inc., Upper Saddle River, N.J.

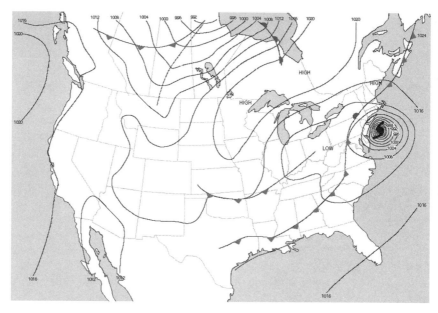

Fig. 14.7. Synoptic chart showing the position of Hurricane Bob, 19 August 1991, prior to landfall in New England. A summary description of Bob and its impact is given in chapter 18.

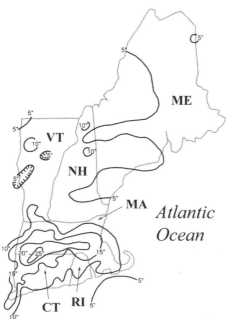

Fig. 14.8. Record New England rainfall totals for August 1955 primarily as a result of Hurricanes Connie and Diane. Revised from Climatological Data for New England, August 1955. From http://airmap.unh.edu/ necaq/Figures/fig3.5.html and *New England's Changing Climate, Weather and Air Quality*, Climate Change Research Center, Institute for the Study of Earth, Oceans, and Space, University of New Hampshire, Durham, New Hampshire, 03824, © 1998 Climate Change Research Center.

Table 14.1
Largest one-day precipitation events
in New England for the sixty-year period
from 1940 to 2000

Location	Date	Rainfall (in.)
Westfield, Mass.	19 Aug. 1955	18.15
Portland, Maine	21 Oct. 1996	11.71
Cockaponset, R.I.	6 June 1982	10.47
Mt. Mansfield, Vt.	17 Sept. 1999	9.92
Torrington, Conn.	31 Dec. 1948	8.91
Middleton, Mass.	6 Oct. 1962	8.64
Woods Hole, Mass.	3 Sept. 1972	8.55
Norfolk, Conn.	16 Oct. 1955	8.20
Brunswick, Maine	11 Sept. 1954	8.05

of 10 inches at numerous sites in Massachusetts and Connecticut. It was particularly damaging because the storm followed the heavy rains produced by Hurricane Connie in southern New England on 12 to 13 August. As a result of these two storms, the month of August 1955 went into the record books as one of the all-time highest for total monthly precipitation, with values reaching over 25 inches for parts of Massachusetts and Connecticut (fig. 14.8).

The second-greatest single-day rainfall event occurred in late October of 1996 (Keim, 1998). This "continental nor'easter" generated the heaviest rainfall values along the eastern coast of New England from Boston to Portland (fig. 14.9). A frequent phenomenon associated with this type of nor'easter is that an upper-level low becomes cut off from the major westerlies (see fig. 2.9), resulting in a very slow moving to stationary low pressure system both at upper levels and at the surface (Davis et al., 1993). That is exactly what happened in the 1996 storm. Coastal and near-coastal sites in Maine and New Hampshire set all-time records for a one-day rainfall event during this storm. Camp Ellis and Gorham, Maine, recorded storm rainfall totals of 19.2 and 19.0 inches, respectively, over a three-day period. Recurrence interval analysis revealed that the event was in gross excess of a hundred-year storm event between Boston and Portland, and at some locations in Maine, it was close to a five-hundred-year storm event. From a statistical point of view, a storm of this magnitude is expected to occur once every one hundred years or five hundred years, respectively. Impacts from such events include river-basin flooding, loss of potable water supplies when sewage plants become flooded, and road and bridge damage (fig. 14.10).

Five of the eight storms listed in table 14.1 occurred in the months from August to October, suggesting that these events are largely tropical, in the form

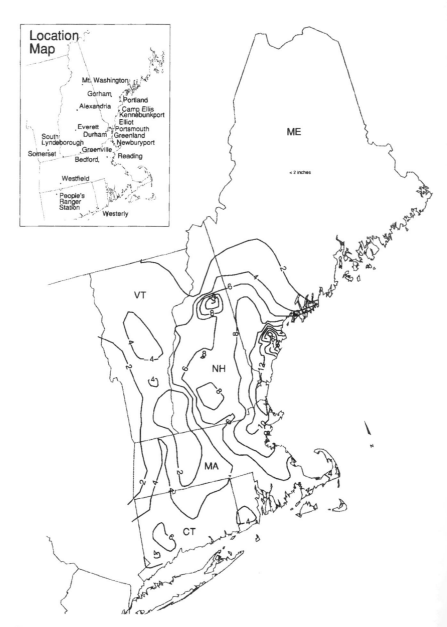

Fig. 14.9. Distribution of precipitation across New England from the 19 to 23 October 1996 storm. Modified from Climate Change Research Center (1998; Fig. 3.6, p. 21, after Keim, 1998, *Bulletin of the American Meteorological Society*), and at http://airmap.unh.edu/necaq/Figures/fig3.6.html. Used with permission from the American Meteorological Society.

Fig. 14.10. Flooding in Durham, New Hampshire, associated with the record-breaking rains of October 1996. Used with permission from Thomas Baker as at http://wintermute.sr.unh.edu/photo/color/events/flood/images/flood7.jpg and http://wintermute.sr.unh.edu/photo/color/events/flood/images/flood5.jpg.

of hurricanes or tropical storms. However, very powerful nor'easters capable of generating heavy rainfall also occur in October (Dolan and Davis 1992). Of the eight events in table 14.1, all but the storms on 6 June 1982, 31 December 1948, and 16 October 1955 had some tropical component. Even the continental nor'easter on 21 October 1996 had some contribution of its moisture transported by means of Hurricane Lili, which was located in the Atlantic at the time. Interaction between tropical storms or hurricanes and mid-latitude storm systems is not unprecedented in New England; the Vermont flood of 1927 was produced under similar circumstances. Though not included in table 14.1 because it occurred more than 60 years ago, this event produced 9.65 inches of rain at Somerset, Vermont, with estimates of 15 inches at higher elevations nearby (Ludlum, 1976; 1996). Other significant New England rainstorms beyond the last fifty years include the 8.00-inch soaking in New Durham, New Hampshire, on 16 September 1932, and 12.25 inches of rainfall reported during a hurricane at Canton, Connecticut on 4 October 1869.

An interesting question concerning these large rain events is whether any global warming could increase the number of such events in the future. Based on this table alone, there is no suggestion that the heaviest of rainfall events are increasing as a result of global warming, since half of these events occurred in the earliest part (1948–1955) of the time period under examination. However, the 1990s have experienced some unusual rain and flood events. Portsmouth, New Hampshire, had two events in the late 1990s, one of which exceeded a hundred-year event (October 1996) and the other exceeded a fifty-year event (June 1998). These results do raise questions about whether this type of rainstorm will be more frequent in the future.

Summary

The two aspects of the hydrological cycle that exert the greatest impact on society and the environment are droughts and flooding rains. Although New England is humid, drought conditions have existed at times over the last 100-plus years. None have been worse than the mid-1960s, when severe drought conditions persisted from 1964 to 1966, peaking in 1965. This drought affected the entire region. However, 2001 surpassed 1965 as the driest year on record from a total precipitation perspective across northern New England, especially in Maine. The duration of the 2001 drought is unknown at the time of this writing. On the other hand, specific events that produce tremendous amounts of rainfall can lead to severe and devastating flooding. Most severe rain storms in New England are associated with tropical systems, with the greatest one-day event being the 18.15 inches falling on 19 August 1955 in Westfield, Massachusetts, from Hurricane Diane. The second-highest single-day rainfall total was the

11.71 inches that fell in Portland, Maine, on 21 October 1996. In this case, a cut-off low pressure system stalled south of the region, leading to the continual pumping of moisture into southwestern Maine, eastern New Hampshire, and northeastern Massachusetts. This storm may have interacted with Hurricane Lili, adding to the precipitation totals, and giving an example of a second method to produce exceptionally heavy rains in New England.

Fig. 15.1. Satellite image of the very powerful nor'easter in the Gulf of Maine on 20 to 21 January 2000. As of the time of this image, the impact was felt primarily in northern New England. Although snow totals were not very high from this storm, exceptionally high winds (sustained 30 to 40 mph) caused severe "white-out" conditions in Down East Maine from blowing and drifting snow. This was a very intense storm, category 5 on the Zielinski winter storm/nor'easter classification scheme (Zielinski, 2002; see text for discussion). Courtesy of Hendricus Lulofs.

Chapter 15

Nor'easters

An old ring-tail snorter of a snow storm. —GEORGE LANG,
STRATHAM, NEW HAMPSHIRE, FEBRUARY, 1893

Of all the weather events that affect New England, nor'easters arguably are the most memorable type of event in the annals of New England weather. This is not surprising, since winter may hold New England in its grip for five to six months of the year. In fact, as we mentioned in the introduction to this part of the book, two of the top ten weather events of the century were blizzards that had a significant impact on New England. These include the Blizzard of '78 and the 1993 Superstorm. Granted, the 1993 Superstorm had a tremendous impact on the entire East Coast, but it does show how these winter storms can wreak havoc on New England culture, the type of havoc that leads to tales that get passed down from generation to generation.

Anatomy of a Nor'easter

Nor'easters are very complex, potentially very large storms that move up the Atlantic seaboard primarily between September and May. Only occasionally do these coastal storms occur in June and August and they almost never occur in July. As these are often very intense low pressure systems, air flow is converging around the center in a counterclockwise motion. Since the classic nor'easter is coastal, the west side of the cylcone has a northeasterly flow, hence the name nor'easter for short (figs. 2.14 and 15.1). In this chapter, we focus on snow-producing nor'easters occuring between November and April, inclusive, although occasionally an October or May nor'easter can produce some snow in New England. We will elaborate on the complex interacting processes responsible for the formation of nor'easters, followed by specific details on the type of damage that they inflict on the region. We then discuss hallmark nor'easters.

We frequently use the term "blizzard" in our discussion, but not all nor'easters are blizzards. A blizzard is defined by several criteria, although not all blizzards meet each one of these. A blizzard is a snowstorm with winds of 35 miles per hour (30 knots) or greater and sufficient snow to reduce visibility to less than ¼ mile (Glickman, 2000). Past criteria have included significant accumulation (usually 10 inches) and possibly temperatures of 20°F or less (Schneider, 1996). Since most large, intense nor'easters often meet these criteria, they frequently are classified as blizzards.

Several factors come into play to form a nor'easter, as Kocin and Uccellini present in wonderful detail in their book *Snowstorms Along the Northeastern Coast of the United States* (1990). We summarize those factors here. The first ingredient is often the formation of a storm around Colorado or in the western Gulf of Mexico. These storms move up the East Coast following a track either on the west side of the Appalachians and through the Saint Lawrence River

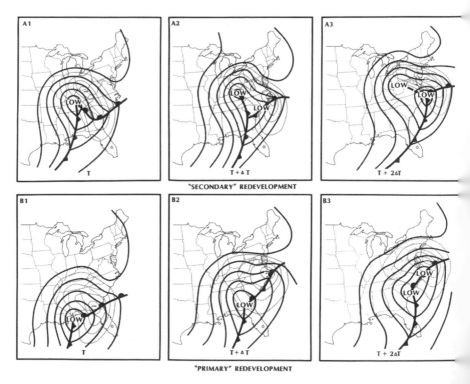

Fig. 15.2. Two of the most common modes of nor'easter development. The top figure shows the scenario whereby a storm system moving across the Ohio Valley region (*A1*) will weaken as a major coast storm forms (*A2*) off the coast (*A3*; that is, "secondary redevelopment"). The bottom figure shows whe a storm that forms in the Gulf of Mexico (*B1*) intensifies (*B2*) as it moves up the eastern seaboard (*B3*; that is, "primary redevelopment"). From Kocin and Uccellini (1990; fig. 8, p. 25). Used with permission from th American Meteorological Society.

valley or on the east side of the mountains along the coast. As those storms approach New England, several things can happen. In some cases, the temperature contrast between the air over land, especially when snow covered, and air originating in the Gulf of Mexico or the Atlantic Ocean is great enough that these storms stay intact, intensify, and become full-blown nor'easters (fig. 15.2). We will discuss the importance of the track taken a little bit later.

The second possibility that can occur as these storms approach the East Coast is that a new low pressure system will form, very often off the coast of North Carolina or maybe slightly further north near the Delmarva peninsula and New Jersey (fig. 15.2). The fuel for cyclogenesis, or storm formation, occurring off the North Carolina coast is the warm water of the Gulf Stream. Air above this water is markedly warmer than the air found over the land of the eastern United States, thus a large energy source is available for storm formation. Often the original low pressure system in this scenario tracks toward the Middle Atlantic states and New England through the Ohio River valley. This storm frequently dissipates and the energy in that system is transferred to the new storm off the Atlantic coastline.

A slightly different scenario can occur along the East Coast once the original low pressure system approaches the coastline, although the final product is the same. As the storm moves up the Atlantic coast, it will appear to re-form at a position slightly ahead of the path it is on. It is almost as if the initial low pressure system "jumped" to a new position along its path while at the same time intensiflying. This "jump" often occurs around the North Carolina coast, with the end product being the formation of a very strong coastal storm.

Occasionally, an Alberta Clipper will move through New England or possibly further south through the Middle Atlantic states with the deposition of only a few inches of snow in parts of New England (see chapter 6 for discussion on this type of storm). Southern to central New England is often affected by such a storm. However, once the storm moves over the warmer ocean waters, it can intensify into a strong nor'easter. The problem for snow lovers is that the intensification may occur too far out to sea to bring heavy snow to much of New England, although Down East Maine may receive heavy snow from this type of setting. Most of the time, major nor'easters that impact New England originate from one of the previous scenarios described.

In many cases, an additional ingredient is needed to ensure that the nor'-easter will be a snow producer, at least somewhere in New England. The presence of a high pressure system north of New England, with its clockwise flow, will bring cold winds from the northeast into New England. When found at all levels of the atmosphere, this cold air keeps the precipitation completely as snow throughout the entire period of the storm.

At the same time, this dome of cold, dense air can inhibit snowfall in New England. As the precipitation from the storm encroaches onto New England, it will run into this very cold, dry air mass and evaporate before hitting the ground. When this occurs, satellite images shown on local television stations and on the Weather Channel will indicate that precipitation is falling over parts of New England, but none of it is reaching the surface. If the high pressure system is too strong, as well as being cold and dry, it may prevent the storm from moving up the eastern coast of the United States, preventing a large snowfall event in northern New England or possibly over the entire region.

Another possibility is that the storm may move out to sea before moving far enough north to affect New England. This is why the Middle Atlantic states may receive much larger snowfall totals out of a single storm than does New England. A depressing thought for snow lovers and skiers! The 6 to 8 January 1996 storm is a perfect example of this scenario: Well over 30 inches of snow blanketed many parts of the Middle Atlantic states, shutting down the U.S. government in Washington, D.C., for several days. Although southern New England was hit hard by this event, with snowfall totals in the 15- to 25-inch range, most of northern New England received less than 5 inches in that storm. Only the hills of southwestern New Hampshire were an exception, receiving 10 to 15 inches. Great Barrington, Massachusetts, recorded 32 inches of snow during this event, while Caribou, Maine, recorded only 0.4 inches.

Once a coastal storm forms, four other important components are essential for creating an intense storm. One factor is the pressure gradient away from the storm. The pressure gradient is the rate of change in pressure between the storm and the adjacent high pressure system. This could be a high located to the north of New England or a high pressure system that builds in from Canada behind the storm. The greater the pressure difference, the stronger the winds, the greater the potential to bring moisture onto the continent from the ocean. This produces high snowfall totals. Pressure differences may be on the order of 0.7 inches of mercury (Hg) per mile (40 mb/km). An exceptionally high gradient occurred during the Blizzard of '78 with a pressure of 29.12 inches Hg (986 mb) for the central low and 31.00 inches Hg (1050 mb) for the adjacent high, thus a gradient of 1.2 inches Hg per mile (64 millibars per kilometer).

The second important aspect of a storm is how quickly the central low pressure intensifies (that is, becomes lower). The stronger the storm, the more the air converges at the surface and rises in the middle of the low pressure center. When the pressure falls very fast, a process referred to as deepening, the storm continues to intensify and pressure gradients increase. This increases wind speeds and the intensity of snowfall as well as total snowfall amounts. When the central low pressure drops at a rate of 0.3 inches Hg (1 mb per hour or 24 mb over 24 hours), the storm is defined as a meteorological "bomb." The formation of a "bomb" can produce high snowfall totals even if the central low pressure itself is not exceptionally low. This process frequently occurs south of New England, but sometimes "bombing" will occur in the Gulf of Maine.

The third additional component needed to develop a large nor'easter is upper-level support, a parameter often mentioned by local weather forecasters and those on the Weather Channel. Upper air support comes in the form of a very fast-moving jet stream and a large trough over the eastern United States. To put it simply, these two conditions are responsible for moving the rising air originating within the surface low pressure rapidly away from the point where it moved into the altitude of the upper-air patterns. The average altitude of the polar jet stream is between 6 and 9 miles above Earth's surface. When this rising air evacuates quickly, more air can converge into the low pressure center at the surface, rise, and force the central pressure of the cyclone even lower. This ongoing pattern produces a very strong coastal storm.

The final ingredient important in determining snowfall amounts is the forward speed of the storm. When a coastal storm "hangs around" for two or more days, snowfall totals are likely to rise to 30 or more inches. Very intense storms that move quickly (that is, that remain in New England for only 12 to 24 hours) may produce only 10 to 15 inches of snow on average over the region. These relatively moderate snow amounts may occur despite a very intense storm. The classic example is the superstorm of March 1993. That was an exceptional storm, but it moved through New England rather quickly, thus snowfall totals across much of the region were in the 10- to 15-inch range. The storm also tracked slightly inland of the New England coastline, producing periods of sleet along coastal areas, which reduced total snow accumulations. We discuss the importance of storm track below. By contrast, the 25 to 27 February 1969 storm was not an exceptionally strong storm, but it "hung around" for several days, producing upward of 40 inches in many places in northern New England. It is often referred to as the "100-hour" storm. We provide more details on these storms later in this chapter.

Given that we now have all the ingredients in place for a major nor'easter, the final critical factor is the all-important storm track. The storm track determines precipitation type across New England: all snow, snow changing to mixed precipitation and possibly to rain, mixed precipitation changing to rain, or all rain. The easiest way to predict the type of precipitation at any one spot in New England is to think of the west side of the storm as the cold side (with movement of air over the cold continent) and the east side of the storm as the warm side (with the movement of air over the warm ocean; fig. 15.3). The final outcome is governed by several other factors, but this is the simplest approach to take. Consequently, a storm moving up central New England or into western New England will bring mostly snow to western areas of the region and mostly rain to coastal areas. Between the two extremes is the "slop," the area of mixed snow, sleet, freezing rain, and rain. Storms moving just off the coast generally will bring all snow to New England, except possibly to Cape Cod, and particularly to the outer Cape. Cape Cod may fall into

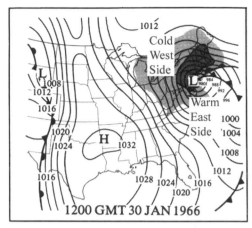

1200 GMT 30 JAN 1966

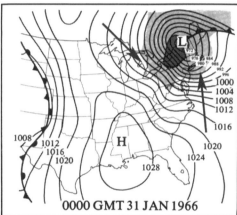

0000 GMT 31 JAN 1966

Fig. 15.3. Position of low pressure center relative to New England as a means of determining the type of winter precipitation that will fall across the region. In this particular example from January 1966, New England is initially on the north side of the storm (*top*), thus the precipitation first fell as snow. However, the storm moved up the Hudson River valley to the west of New England (*bottom*). Most of the snow turned to rain when most of New England ended up on the warm, east side of the storm with air flow coming from the south (as shown by arrow). Areas to the west of the low would receive snow on cold northeast to northwest winds (as shown by arrow). Modified from Kocin and Uccellini (1990; fig. 63, p. 151). Used with permission from the American Meteorological Society.

the mixed-precipitation or rain regime. As one can see, the trickiest scenario is the storm that moves almost directly along the coastline. In this situation, exactly where the mixed precipitation will fall and where it will all turn to rain is rather complex and difficult to forecast. The strength of the storm is also important, as a very intense storm can move the warmer ocean air further inland then a weaker storm can. The mixed-precipitation zone would then occur further inland despite a storm track that was just east of the coastline. The entire scenario is so complex with these winter coastal storms that a storm track difference of only a few miles to tens of miles can make a very large difference in snowfall totals. It is a very hard job to predict where these boundaries will occur, and meteorologists in New England probably have one of the toughest jobs in the United States when it comes to predicting snowfall amounts associated with a major nor'easter.

In addition, one should not forget that once a storm moves north of the region, the winds will be coming around the west side of the storm with support from a cold high pressure system to the west of New England. This flow can change the precipitation on the rain side of the storm back to snow. In fact, areas between the exiting cyclone and the incoming high pressure center can be quite unstable. This instability can lead to another couple of inches of snow at the tail end of a storm or possibly frequent snow squalls during the day or two after the storm. These squalls easily can add an additional few inches to the total amount of the snow event. Snow squalls are particularly prevalent in the interior mountains and less so along the coast. As a result, the complete sequence for a coastal storm may vary from all snow at one extreme, to snow and mixed precipitation or rain and back to snow, to mixed precipitation to rain and then to snow, or to all rain, at the other extreme, with some snow flurries in the days following the storm. This wide variety of conditions associated with a single nor'easter is a microcosm of New England's weather and climate.

Nor'easter Impact

The actual impact of a nor'easter also may be quite variable. Large amounts of snow, especially when accompanied by strong winds and drifting, may block roadways and cause extensive delays at airports. Zielinski (2002) developed a scheme to classify winter storms by their intensity much like hurricanes and tornadoes (see chapters 18 and 17). Using three factors that determine how strong a storm may be—central low pressure, maximum pressure gradient, and rate of deepening—the intensity index of a nor'easter will fit into specific categories on a 1 to 5 scale (table 15.1). The potential impact per category is given in table 15.1. In addition, a duration factor is added to the intensity index. These numbers can then be used to assess the possible impact of a nor'easter. In addition, mixed precipitation often creates treacherous travel coditions and downed power lines and trees. We discuss these risks in the next chapter on ice storms. Finally, the great flow of wind off the ocean and accompanying high waves can cause extensive beach erosion and coastal flooding particularly around high tide. The maritime industry is well aware of the dangers to shipping and fishing with large coastal storms. In fact, in the minds of individuals who operate boat touring companies, "nor'easter" is a dirty and unwelcomed word. The fishing and shipping industries do not appreciate frequent nor'easters moving into New England waters. Lobster fishing may be especially hard hit, as fishers may not be able to get out to sea, and their traps can be ripped up by the high waves. "The Perfect Storm" brought to the attention of the public the power of these behemoths, although that was not a winter storm. Nevertheless, the same dangers hold true for winter nor'easters, as evidenced by the

Table 15.1

Zielinski winter storm/nor'easter intensity index categories with duration factor and potential impact

STORM INTENSITY

Category	Intensity index (I)	Intensity rating	Duration factor (DF)	Forward speed range (mph)	Duration factor rating
1	0–24	weak	1	>54	very fast
2	25–48	moderate	2	41–54	fast
3	49–72	strong	3	27–40	moderate
4	73–96	very strong	4	14–26	slow
5	>96	powerful	5	0–13	very slow to stationary

POTENTIAL IMPACT

Category	Maximum snowfall amounts	Maximum snowfall rate	Potential wind speeds	Maximum drifting potential	Closings/delays on communities, schools, and travel	Impact on coastal and maritime interests	Nature of disruption, terminology of Rooney (1967)
1	<10 in	very low <1 in/hr	weak	minor <20 in	Maybe minor (hours)	Minor	Minimal to nuisance
2	10–20+ in	moderate 1+ in/hr	strong	moderate 3 ft	Maybe moderate (hours to 1 day common)	Minor to moderate	Nuisance to Inconvenience
3	20–30+ in	high 2+ in/hr	gale force	high 4–6+ ft	Possibly extensive/lengthy (several days possible)	Moderate to severe	Inconvenience to crippling

4	30–40+ in	very high 2–3+ in/hr	gale force to hurricane	very high 6–10+ ft	Probable extensive/lengthy (up to a week may be common)	Severe	Crippling to paralyzing
5	40–50+ in	overwhelming >3+ in/hr	gale force to hurricane	exceptional 10–15+ ft	Extensive/lengthy (up to a week common)	Extreme	Paralyzing

Categories are presented in the form (I.DF). For example, a 5.4 nor'easter would be a very strong, slow-moving storm, thus it would have the potential for heavy snowfall or rain depending on storm track.

Slow-moving storms may produce snowfall totals greater than that suggested for a particular category.

Storm track will greatly influence the type of precipitation (snow/rain/mixed) and thus the impact. If precipitation type is rain, maximum amounts may be on the order of <1 inch for category 1, 1–2 inches for category 2, 2–3 inches for category 3, 3–4 inches for category 4, and 4–5 inches for category 5, assuming a generalized snow water equivalent of 10 inches snow (as above) = 1 inch water.

Wet snows have a greater impact than dry snows of an equivalent depth (Rooney, 1967).

Storms in the central U.S. may produce lesser amounts of precipitation (and drier snows) than nor'easters of an equivalent category.

Wind speeds over 15 mph have a greater impact during a snowstorm than lower wind speeds (Rooney, 1967).

Wind speeds over open water will tend to be greater than over land for a given category.

Modified from Zielinski (2002, Bulletin of the American Meteorological Society). Used with permission of the American Meteorological Society.

"Portland Storm." This 36-hour storm on 26 to 27 November 1898 led to the sinking of at least 150 ships, including the steamship *Portland*, and an estimated 450 lives lost. In fact, in some ares of New England, such as Scituate, Massachusetts, "the Portland Storm" was the hallmark nor'easter until the Blizzard of '78 (D. Arnold, *Boston Globe*, 23 November 1998).

One of the more interesting coastal impacts of a nor'easter is the possibility of a damaging storm surge. This phenomena is more often associated with hurricanes (see chapter 18), but very strong nor'easters can push large amounts of water through the northeast quadrant of the storm (fig. 15.4). The March 1993 Superstorm produced a damaging storm surge on the western Florida coast as it moved past that region. When this occurs at high tide, the resulting tidal surge can have a major impact along the coast. This is the exact situation that occurred during the "Groundhog's Day Storm" of 1976 (Morrill, 1977). This very powerful nor'easter moved north through western Maine, bringing strong winds (115 mph in Southwest Harbor and 69 mph in Rockland) and high surf to the mid-coast and Down East coasts of Maine. Waves as high as 14 feet were reported. The storm moved onshore at the time of high tide, which compounded the situation. Tidal levels were up to or over 5 feet above expected high tide in many places along the Maine coast, with some areas reaching 10 feet above expected high tide (Morrill, 1977). However, the most extreme consequence of this storm was a result of the confining shape of the Penobscot River embayment in Maine (fig. 15.4). The tremendous tidal surge pushed up the Penobscot River produced water levels of up to 17 feet above mean low tide levels and water depths of 12 feet in the riverfront area of Bangor. The high water moved up Kenduskeag Stream (which flows into the Penobscot in Bangor), flooding about 0.5 square miles of the riverfront area of Bangor. The rapidity of the water rise submerged about two hundred cars and trapped several individuals trying to move their cars from the rising waters.

Hallmark Nor'easters

Despite the many strong nor'easters that have hit New England over the last few hundred years, including "The Portland Storm" and "Groundhog's Day Storm," several stand head and shoulders above the rest. We discuss some of the specific details of these storms to give an indication of how powerful and memorable are these large storms.

The Blizzard of '88. By far the greatest New England snow storm is the blizzard of 12 to 14 March 1888. This storm dropped excessive amounts of snow over much of south-central to central and western New England. Greatest accumulations were the 40-plus inches in southern Connecticut (Middletown received 45 inches and New Haven 42 inches) and in extreme western Vermont

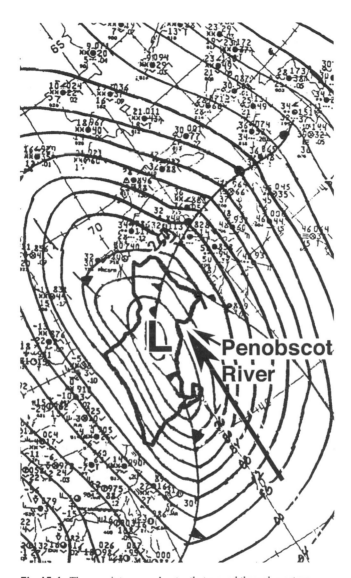

Fig. 15.4. The very intense nor'easter that moved through western Maine on 2 February 1976 at the time of high tide. The intense storm surge that pushed up the Penobscot River embayment on very strong south to southeast winds (large arrow) produced tidal levels over 17 feet above mean low tide in Bangor. The storm surge moved up Kenduskeag Stream, which empties into the Penobscot River in Bangor, flooding a portion of the riverfront in Bangor under 12 feet of water. Weather symbols were removed from within New England for clarity. Modified from Morrill (1977) as graciously provided by Hendricus Lulofs.

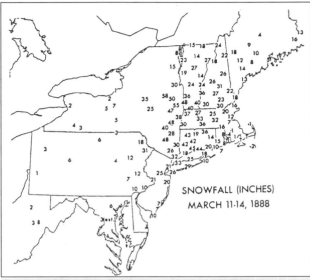

SNOWFALL (INCHES)
MARCH 11-14, 1888

Fig. 15.5. Snowfall amounts resulting from the Blizzard of '88 (*top*). The photograph shows individuals standing atop drifts in front of the American Hotel, Hartford, Connecticut, after the blizzard. Top figure after Kocin (1983; fig. 4, p. 1261, *Bulletin of the American Meteorological Society*), used with permission from the American Meteorological Society. Bottom photo from pg. 448, State Archives, Connecticut State Library, E. P. Kellogg, Hartford, photographer. Used with permission from the Connecticut State Library.

(fig. 15.5; Kocin, 1983). Greatest snowfalls for the storm, however, were the 48 to more than 50 inches that fell around the Albany, New York, area, including the jackpot of 55 inches at Troy. Although the heaviest snowfall amounts were found in eastern New York, much of Vermont, New Hampshire, and western to central Massachusetts received 20 to 30 inches of snow. Those totals would have placed the Blizzard of '88 at the top of the list for largest snowfall events in many locations around New England over the last seventy years, or at the least, within the top five. On the other hand, southeastern New England and particularly coastal sites received only about 5 to 7 inches, either because the turnover from rain to snow at these sites occurred very late in the storm or because the particular site was close to a fluctuating rain/snow line.

The primary nor'easter formed off the North Carolina coast following dissipation of a small to moderate low pressure system that moved across the central part of the country toward the Great Lakes. The new storm moved up the Atlantic seaboard as it intensified. The key component of this storm became the lack of movement once it intensified into a very strong storm (Category 4; Zielinski, 2002). The storm essentially sat over Block Island, Rhode Island, for two days, while the pressure gradient continued to increase as a high pressure system moved across southern Canada from the west. This increased pressure gradient helped move more and more moisture-laden air into western and southern New England. At the same time, more frigid air was moving into western New England, which kept the precipitation as snow. Enough warm air moved into eastern New England that snowfall totals were miniscule compared to the tremendous amounts in Connecticut and eastern New York. The entire area of western Connecticut and eastern New York was shut down as transportation became impossible. Only southeast New England and Maine received less than 20 inches of snow. Parts of western Maine did receive between 10 and 20 inches of snow. Given the exceptionally high snowfall totals and the extensive area covered by at least 20 inches of snow, this storm justifiably is the hallmark New England blizzard.

The Great Snow of 1717. Prior to the Blizzard of '88, the hallmark New England snow event was the Great Snow of 1717 (fig. 15.6). The major difference between these two events was that the snows that accumulated during the 1717 event came from a series of four storms from 27 February to 7 March 1717, rather than the single storm that produced the 1888 blizzard (Ludlum, 1976). The Great Snow of 1717 consisted of two or three major storms (1, 4, and 7 March), each of which produced significant amounts of snow around parts of New England and particularly in eastern Massachusetts. The storm on 27 February was more minor, with a combination of snow, sleet, and rain depending on location. The accumulation from all four storms over this seven-day period appears to have been on the order of 40 inches in Boston and possibly up to

Fig. 15.6. Woodblock print depicting how difficult travel may have been during the Great Snow of 1717. From Ludlum (1976; p. 53).

60 inches north of Boston. Information on these storms comes from accounts in diaries by several individuals in New England, including Joushua Hempstead of New London, Connecticut. The totals noted in the written record rival snow depths in western and southern New England from the March 1888 storm. Drifts during the 1717 event may have been in the 15- to 20-foot range, making passage almost impossible.

The Blizzard of '78. Of all the large snowstorms that hit New England in the last century, the 7 to 8 February 1978 storm probably sticks in the mind of most individuals who experienced it more than any other storm. It is probably most famous for its extreme intensity (snow, wind, and resulting drifts; fig. 15.7), great explosiveness (the drop in central pressure was more than 3 millibars in 3 hours), which surprised many forecasters, and its tremendous impact on the Boston metropolitan area (Kocin and Uccellini, 1990). The large number of people affected by the storm places it among the hallmark storms of New England, and it has been rated as one of the top weather events of the twentieth century in the United States. Snow totals in parts of eastern Massachusetts and southward toward northern Rhode Island exceeded the 30-inch level, with greater than 20-inch totals throughout much of southeast New Hampshire, central Massachusetts, eastern Connecticut, and eastern Vermont (fig. 15.8). An unofficial total of more than 50 inches fell in western Rhode Island. Winds were exceptionally strong during the storm, building drifts of 8 to 10 feet in areas around Boston. Blocked roadways and abandoned cars on the major arteries of the Boston metropolitan area were a result of the tremendous snowfall rates and visibilities near zero during the storm. Coastal flooding and erosion were prevalent along the coast of eastern Massachusetts.

The two major factors contributing to the high snowfall totals and high wind speeds were the great pressure gradient between the central low and the high pressure system north to northwest of the storm center, and the period of slow movement that corresponded with the period of rapid intensification. Hurricane force winds helped push the abundant snow into tremendous drifts that shut down the Boston metropolitan area for a week. An interesting aspect of this storm was that the low pressure itself was not nearly as low as many other large nor'easters over the last few decades. The central pressure only dropped to about 29.12 inches of mercury (986 millibars). Many large and intense nor'easters can have central low pressures around 28.30 inches Hg (960 mb). However, the very large high pressure system west of the storm set up a tremendous pressure gradient, which fueled the flow of air off the ocean, bringing strong winds and deep snows. The impact of the storm may actually have been worse than other hallmark storms because of the advances in modern society since the Blizzard of '88.

We could describe many other extreme snow-producing nor'easters in New England. In table 7.3, we compiled a list of several storms that are found among the top fifteen snow events of the last seventy years for several sites across New England. Certainly, the year of 1969 will go down as having two

Fig. 15.7. Two photographs showing the tremendous drifting associated with the Blizzard of '78. The top photograph shows the blocked doorway of the Road House Restaurant in Marlboro, Massachusetts. The bottom photo shows a buried lightpost in a Marlboro neighborhood. Used with permission from Eric Werme as at http://people.ne. mediaone.net/werme/blizz78. html.

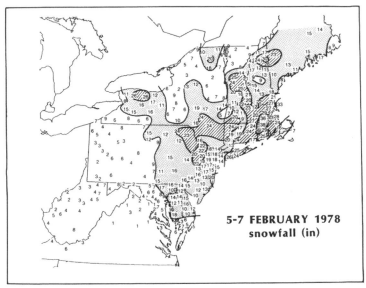

**5-7 FEBRUARY 1978
snowfall (in)**

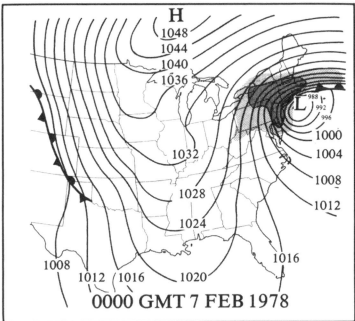

0000 GMT 7 FEB 1978

Fig. 15.8. Snow depths from the Blizzard of '78 (*top*) and the position of the major circulation systems during the height of the storm (*bottom*). From Kocin and Uccellini (1990; fig. 95, p. 213, and fig. 96, p. 214). Used with permission from the American Meteorological Society.

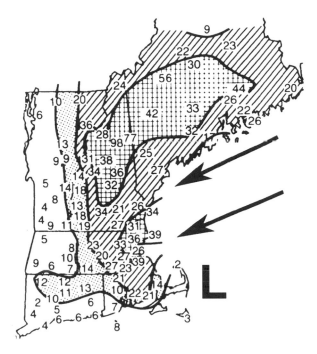

Fig. 15.9. Snow depth from the 22 to 28 February 1969 winter storm showing the tremendous snowfall totals in eastern New England and the dramatic drop-off in totals on the western slopes of the central mountains. The exceptionally high snowfall totals were a direct function of the storm's stationary nature and easterly flow off of the ocean (shown by arrows). The storm remained off of Cape Cod for almost 48 hours, as shown by the L. From Kocin and Uccellini (1990; fig. 78, p. 179). Used with permission from the American Meteorological Society.

noteworthy snow storms within it, the 27 to 29 February event and the 24 to 26 December storm. In both cases, more than 20 inches of snow fell across a wide area of New England. In fact, several totals for the 25 to 29 February event are still records, such as the 77 inches at Pinkham Notch, New Hampshire, and the 99 inches on Mount Washington (fig. 15.9). Forty-plus to 50-plus inches also accumulated in northwestern Maine. The Superstorm of 1993 also produced a high amount of snow across the region, although totals were not that excessive compared to the Blizzard of '78 and the 1969 snow events. The 1993 storm was exceptionally powerful, but it moved through New England rather quickly, which kept snowfall totals down. Furthermore, some sleet mixed in along coastal areas.

More recently, the 6 to 8 March 2001 storm proved to be quite a focus of attention not only for the people of New England, but for much of the eastern coast of the United States. Winter storm watches were in place for several days before this major nor'easter formed, initially for the area around Washington, D.C., and Philadelphia. As the system began to develop, meteorologists predicted that the heaviest snowfall would be in the northern Pennsylvania to southern New York and New York City area. However, the complex nature of these systems won out, and the highest snowfall totals eventually were recorded in southeastern New Hampshire. Maximum snow reported from this storm was 50 inches at Nottingham, New Hampshire, while 32 inches was "unofficially" reported to the National Weather Service for Durham, New Hampshire. The 32 inches in Durham makes this the all-time biggest snowfall on record. Barry Keim reported 23 inches in Barrington (fig. 15.10). This storm helped to produce a record snow depth of 38 inches for Greenland, New Hampshire, on 10 March, compared to an average maximum snow depth of about 18 inches. The reason for such a deep snowpack was that much of New England had no significant thawing during the winter of 2000/2001 and the snowcover was persistent from mid-December through early April.

Several other New England storms produced widespread snowfall totals, but individual totals are not excessive. Essentially, many of these storms followed a track that allowed snow to fall over much of New England. For example, the 5 to 7 April 1982 storm was a powerful but fast-moving system, thus total snowfall amounts were generally in the 10- to 15-inch range. However, this storm highlights the potential for large late-season snows that feed off the cold temperatures still existing over land and the encroaching warm air from the south. The record-setting "April Fools' Day" storm on 1 April 1997, with a total of 25.4 inches in 24 hours at Boston, also shows the potential for late-season monsters. Come April, New Englanders expect that mud season is rapidly approaching and winter is ending. Such thoughts can easily be crushed with one of these large April nor'easters.

Summary

Nor'easters are possibly the most prevalent severe weather events that affect New England. These large coastal storms with their counterclockwise flow around the low from the northeast may occur anytime of the year, but they are most abundant from October to April. They are rare in the heart of summer. Consequently, the notoriety of nor'easters comes from their production of large snows, high winds and associated drifting, and coastal erosion and flooding from high waves and tidal surges. Several key ingredients are essential in the formation of a strong nor'easter. A strong temperature contrast must exist, such as the contrast of air over the warm Gulf Stream waters off the Atlantic Coast with air over the frigid continent, high pressure to the north to keep the cool northeast flow in New England (for a snow-producing nor'easter), and a strong jet stream to assist in convective processes and to lower surface pressure. A nor'easter may originate in the Gulf of Mexico and later intensify off the North Carolina coast or may originate from a storm that moved across the country while weakening in the Ohio River valley. The energy from that storm often contributes to the secondary nor'easter that forms off the coast. These conditions all contributed to the formation of several very large nor'easters

Fig. 15.10. Heavy snows in Barrington, New Hampshire, associated with the 6 to 8 March 2001 blizzard. Photos by Barry Keim (*left*) and Ellen Keim (*right*).

such as the Blizzard of 1888 (40-plus inches in Connecticut), the 1717 Great Snows, and the infamous Blizzard of '78 that completely shut down the Boston metropolitan area. Many other storms could fall into the "hallmark" category, but these are some of the more famous storms to New Englanders. Regardless of which storm is talked about or when the next "really big" winter storm arrives, stories will continually be passed down about the famous "Blizzard of . . ."

Chapter 16

Ice Storms

*. . . we should still have to credit the weather with one feature
which compensates for its bullying vagaries—the ice storm.*
—MARK TWAIN

N ew England experiences almost all types of severe weather, but ice
storms have the potential to be one of the most devastating. We can
present no better evidence of the processes that cause an ice storm to
develop and of the damage such a storm can cause than the ice storm of 5 to
9 January 1998, quite possibly the worst ice storm on record in North America.
The storm affected northern New England, upstate New York, southern Que-
bec, Ontario, and New Brunswick, Canada. In this chapter, we will use the
January 1998 storm to illustrate how an ice storm forms and what its impacts
on society may be. New England history records other intense ice storms that
caused impressive damage; we will discuss these hallmark storms as well.

Anatomy of an Ice Storm

Freezing rain and glaze both refer to processes contributing to ice storms, but
it takes a major freezing event with excessive amounts of ice accumulation
on trees and power lines to be termed appropriately as an ice storm. Freezing
rain typically occurs under conditions of a temperature inversion, whereby the
lower portion of the atmosphere contains a relatively warm, saturated layer
with a cold layer right at the surface and a cold layer above the warm layer
(fig. 16.1). Under these conditions in non-summer months, the warm layer may
contain temperatures above freezing while the surrounding layers are below
freezing. These conditions are ideal for the formation of freezing rain and possibly an ice storm.

　　Within a cloud, snowflakes gradually increase in size and weight until they
fall out of suspension from within the cloud and are pulled to Earth's surface
by gravity. In the event of a temperature inversion, upon passing through the

warm layer along this downward trajectory, the snowflakes melt into liquid water, which is again chilled below freezing in the cooler portion of the atmosphere below. When this layer below freezing at Earth's surface is thick and the raindrops are brought well below freezing temperatures, sleet is typically produced (fig. 16.1). However, when this cold layer at the surface is thin and raindrop temperatures are only slightly below freezing, the drops typically do not freeze in the air, but continue to fall as supercooled liquid drops that freeze upon contact with a surface. This type of precipitation is called freezing rain. The build-up of ice on the surface is called glaze. No definitive amount of ice accumulation has been specified to warrant the designation of an ice storm.

Data from New Hampshire for the 1998 ice storm clearly demonstrate the processes that cause freezing rain. For example, the summit of Mount Washington penetrated the inversion layer where temperatures mostly remained above freezing during the icing event, even climbing as high as 45°F. This was

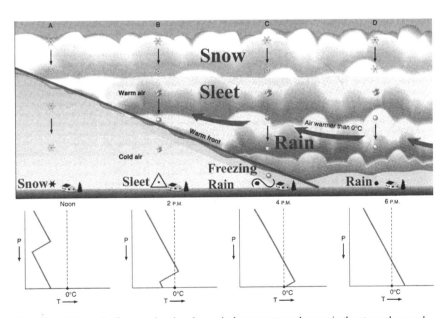

Fig. 16.1. Schematic diagram showing the vertical temperature changes in the atmosphere and resulting precipitation type associated with a warm front advancing over colder air during the winter when below freezing temperatures exist. Sections A to D of the figure depict the changing conditions at a single location, noted by the house and tree on the surface, over the 6 hours that the warm front advanced over this spot. The thickness of the below-freezing layer through which the melted precipitation must pass is the primary control of whether sleet or freezing rain occurs at the surface (*sections B and C*). The same diagram could be used to show changes in space at a single time as the warm front advances from right to left. In this latter scenario, it would be snowing at the leftmost house at the same time it is raining at the rightmost house. Modified from Danielson et al. (1998; fig. 7.16, p. 198). Used with permission from The McGraw-Hill Companies.

Table 16.1

Concord, New Hampshire; observations at 3-hour intervals

			Temperature °F				Wind	
Hour (LST)	Sky Cover	Weather	Dry Bulb	Dew Point	Wet Bulb	Relative Humidity (PCT)	Speed (MPH)	Directio (Tens o degrees
JANUARY 7								
01	ovc	Mist	34	33	34	97	0	00
04	ovc	Mist	33	32	33	96	6	06
07	ovc	Mist	33	32	33	96	5	12
10	ovc	Light freezing rain and mist	32	30	31	92	7	10
13	ovc	Light freezing rain and mist	32	30	31	92	6	04
16	ovc	Light freezing rain and mist	32	31	32	96	6	03
19	ovc	Light freezing rain	32	31	32	96	5	05
22	ovc	Low drifting mist	33	32	33	96	5	03
JANUARY 8								
01	ovc	Heavy rain and mist	34	32	33	92	5	36
04	ovc	Rain and mist	33	32	33	96	7	02
07	ovc	Light rain and mist	34	32	33	92	6	05
10	ovc	Light rain and mist	35	33	34	93	7	02
13	ovc	Heavy rain and mist	35	33	34	93	3	02
16	ovc	Light rain and mist	34	32	33	92	3	03
19	ovc	Light rain and mist	34	32	33	92	0	00
22	ovc	Mist	33	32	33	96	3	29

Three-hour weather observations at Concord, New Hampshire, on 7 to 8 January during the 1998 ic storm. Dry bulb is a measure of the ambient air temperature, while the wet bulb is a measure of the temperature while considering evaporation off the thermometer bulb, yielding information about the humidity. Weather conditions, shown in the third column, clearly change from rain to freezing rain and back again as the temperature (both dry and wet bulb) goes from above 32°F to 32°F or below, and ba to above 32°F. Wind directions are consistently between north (00 degrees) and northeast (45 degrees).

the highest January temperature ever recorded at the summit. In contrast, areas at lower elevations were colder. The locales at lower elevation that had temperatures consistently below 27°F received sleet, those that had temperatures between 27 and 32°F received freezing rain, and those areas that had temperatures above 32°F received rain. Even shifts in temperature of 1°F (from 32 to 33°F) meant the difference between glaze and plain old rain. This explains why the damage from the 1998 storm was so spatially scattered within short distances. Damage also was focused on the northeast slope of many hillsides, reflecting the prevalence of north to northeast wind during the entire event.

Since freezing rain and sleet occur under such limited conditions, these events are difficult to predict and cause serious problems for weather forecasters. This only exacerbates the potential hazard, because most people are caught by surprise when ice storms hit an area. Given the dynamic changes within the atmosphere over short durations, some empirical studies have found that precipitation can fluctuate between rain, snow, ice, and back in as little as 30 minutes (Williams, 1992). One prime example of these rapid shifts in conditions is provided by the three-hour observations at Concord, New Hampshire, on 7 to 8 January 1998 (table 16.1). When the temperatures were 33°F or above, mist, rain, and drizzle were reported; however, when the temperature dropped by 1°F to 32°F, conditions shifted to freezing rain.

Despite the specific conditions necessary for ice storm formation, these events occur somewhere in New England almost annually. Sometimes glaze forms in a narrow band between areas of snow and rain, or it can be widespread when the appropriate meteorologic conditions exist over a vast area, as happened in early January 1998. Glaze is also common in mountainous terrain, because of changes in temperature with elevation, even when rain occurs in the lowlands. New England is prone to such events given its tremendous relief, ranging from sea level along its coastal zone up to 6,262 feet atop Mount Washington, with major portions of the region consisting of hills and mountains. In fact, valleys within the hills and mountains of New England are ideal places for cold air to be trapped at the surface, and thus one of the necessary ingredients for freezing rain is easily attained in this region. These conditions bring great variety in regional and local weather conditions. Proximity to the ocean also can increase the likelihood of experiencing an ice storm, because the flow of warmer air off the ocean can center in mid-levels of the atmosphere around the low pressure center. This flow creates temperature profiles conducive to ice storms near the coast. At the same time, colder conditions throughout all levels of the atmosphere may exist further inland, where the precipitation will fall primarily as snow.

When a major event occurs, such as the January 1998 ice storm, some individuals may want to attribute the cause to a specific factor, particularly global warming. The set of conditions that led to that ice storm could potentially develop any winter, thus there is no reason to say that we will have more of these types of events in the future with global warming. It is equally incorrect to say that we will have more ice storms and less snow with a warmer climate. In fact, many winter storms have the potential to produce freezing rain and sleet as they move across New England. Some of these storms may produce more than others, but freezing rain is a significant component of our weather and climate system in New England.

Ice Storm Impact

Society at large is affected by ice storm events in a number of ways. Freezing rain events also can be dangerous and destructive, just to a lesser degree overall. Streets can be covered by glaze in minutes, reducing driving conditions to dangerous levels. Automobile accidents are frequent, and occasionally lead to fatalities. Sidewalks are similarly affected, leading to unsafe conditions for urban pedestrian traffic. This can be particularly problematic for the elderly who are susceptible to broken bones. Glaze also builds up on tree limbs and power lines, causing loss of power through the direct collapse of power lines or by tree limbs falling onto power lines (figs. 16.2 and 16.3). In some cases, the loss of power may result in the inability to heat one's house, a potentially dangerous situation during the winter months, particularly for the elderly and sickly.

Of all of these impacts, glaze accumulation on trees may very well pose the greatest economic damage (table 16.2). In January of 1998, for instance, an estimated 7.1 million hectares of forest were damaged in upstate New York and northern New England (DeGaetano, 2000). Costs for the maple sugaring industry were especially high. Not only can forest ecosystems be damaged, but urban trees are frequently destroyed (Rebertus et al., 1997; Sisinni et al., 1995). Glaze accumulations on trees and power lines can range from a trace to over an inch in thickness and can increase the weight of tree branches to over thirty times their normal weight (Hauer et al., 1993). This weight easily can bend birch trees completely to the ground and brittle pine trees snap off at their tops. Power lines are frequently pulled to the ground by falling trees in rural and urban areas, creating dangers of electrocution, while leaving major urban areas without power for days. Some rural areas may go for weeks without electricity. Following the January 1998 ice storm, over 600,000 customers in the northeastern United States were without electricity, which translates to about 1.4 million people. Even more importantly, this disruption indirectly resulted in seventeen deaths (DeGaetano, 2000). Ice storms indeed have the potential to be a highly destructive power of nature (table 16.2).

Fig. 16.2. Photographs showing the several inches of ice that accumulated during the 1998 ice storm on grass (*top*), trees and power lines (*bottom*), and the resulting tree damage (*bottom*). Pictures from Maine. Top photo by John Jensenius, bottom photo by Hendricus Lulofs. Photos from National Weather Service forecasting office, Gray, Maine, web page (http://www.nws.noaa.gov/er/gyx/icepix/ice_storm_98.htm).

Fig. 16.3. Photographs showing ice accumulation on power lines that led to broken poles in Maine (*top*). Photos by Thomas Berman. Photos from National Weather Service forecasting office, Gray, Maine, at http://www.nws.noaa.gov/er/gyx/icepix/ice_storm_98.htm.

Table 16.2
Impact of the 1998 ice storm on New England

	Maine	New Hampshire	Vermont
ECONOMIC LOSS (MILLIONS OF DOLLARS)			
Electric utilities	58.0	4.0	6.9
Communications	24.7	not given	not given
Insured losses	55.0	15.0	10.0
Insurance deductibles	10.1	2.7	2.2
Agriculture			
Total	25.0	not given	not given
Dairy	not given	not given	3.0
Maple syrup	not given	1.2	4.8
Forests	31.3	8.6	5.8
Family Assistance	8.0	not given	not given
INSURANCE CLAIMS (THOUSANDS)			
Personal	33.0	7.8	6.9
Commercial	7.3	3.0	1.9
ELECTRIC CUSTOMERS LOSING POWER (THOUSANDS)			
	338.0	68.0	60.4

From Degaetano (2000, *Bulletin of the American Meteorological Society*). Used with permission from the American Meteorological Society.

Hallmark Ice Storms

The January 1998 event is truly the hallmark New England ice storm. As we used it to provide examples about ice storm origin, processes, and impact, we will not elaborate on it in this section. However, compilations by Ludlum (1976) show that several other major ice storms have struck various parts of New England over the past two centuries. Although several of these have produced ice accumulations on par with the 1998 storm, these other storms affected significantly smaller areas. Prior to 1998, perhaps the worst ice storm devastation occurred during the 26 to 29 November 1921 storm (fig. 16.4). In this instance, seventy-five hours of rain, freezing rain, sleet, and snow fell over central and eastern Massachusetts, producing slightly over 4 inches of mixed precipitation in Worcester, Massachusetts. Ice was 2 inches thick on many power lines, and over 100,000 trees were damaged.

In December 1973, a thirty-six-hour period of mixed precipitation and freezing rain produced total precipitation amounts between 1 and 3 inches in parts of Connecticut. Almost one-third of the state was without power, and tree

Fig. 16.4. Ice-coverd phone lines in the November 1921 ice storm in Worcester, Massachusetts (*top*), and ice-covered trees on Richards Avenue, Portsmouth, New Hampshire, in the 28 to 29 January 1886 ice storm. Top photo from Ludlum (1976; p. 58). Bottom photo from the Portsmouth Public Library as at http://www.seacoastnh.com/earlyphotos/icestorm/index.html. Used with permission from J. Dennis Robinson at SeacoastNH.com.

damage was estimated to be greater than that produced by the 1938 hurricane. Other storms throughout New England in the late 1800s (fig. 16.4) and early 1900s are known to have produced large amounts of mixed precipitation, such as the 7 inches of sleet and ice in northwest Connecticut in February 1898. But these storms were all much smaller in scope than the hallmark ice storm of 1998. Furthermore, the increase in technical advances of society in recent times—phone lines, cable TV, electricity—has led to much greater damage and disturbance of lifestyle with this type of storm, because we have become so dependent on these technologies. Interestingly, in colonial times, travel would have been inhibited to some degree, but the impact probably was not nearly as severe as that from present-day ice storms.

Summary

Ice storms are things of both beauty and of destruction; they are "nature's beauty and the beast." Even the glaze associated with minor freezing rain events can produce very shiny and sparkling views and scenery, while at the same time causing traffic accidents and personal injuries from falls on slippery surfaces. The thick accumulation on surfaces such as trees and power lines that comes with lengthy periods of freezing rain are the storms that legends are made of. The hallmark ice storm for New England is, without doubt, the January 1998 ice storm. Other New England ice storms have produced equal amounts of damage, but these other storms affected much less area than did the 1998 storm. Freezing rain occurs when a layer of warm air is sandwiched between two layers of below freezing air. Snow melts as it falls through the warm layer. The thickness of the freezing layer beneath determines whether sleet falls at the surface (a thicker cold layer) or if freezing rain occurs (a thin cold layer that causes the supercooled liquid droplet to freeze upon contact). These conditions inevitably occur every winter in New England. The length of time that the freezing rain falls determines whether a thin glaze forms on surfaces or a thick accumulation of ice occurs, meriting the designation of ice storm.

Chapter 17

Tornadoes

The sky was turning green and black. It was horribly menacing.
—CANDACE FATEMI, NORTH WINDSOR, CONNECTICUT,
3 OCTOBER 1979

Tornadoes are arguably the most violent storms on Earth (fig. 17.1). They can have windspeeds well over 250 miles per hour, with documented examples of trucks and railroad cars being lifted off of the ground and dropped hundreds of feet away (Lutgens and Tarbuck, 1998). Large trees are easily uprooted and houses present little resistance to the most powerful tornadoes. Although Mount Washington holds the record for the highest wind speed ever measured, it is believed that many tornadoes in the United States have higher winds than this. The fact remains that tornadic winds have not been accurately measured. There are two reasons for this. First, it would be rare for a tornado, which occupies only a small area, to move over a pre-existing anemometer—an instrument that measures wind speed. Second, if a tornado did track over an anemometer, especially one attached to a mobile unit put into place by "storm chasers," it most likely would damage the instrument because most sensors are not designed to handle conditions that violent. Mount Washington has special instrumentation that is designed to cope with the windy conditions that prevail on the mountain.

Anatomy of a Tornado

Tornadoes form in association with thunderstorms, usually imbedded within a mid-latitude cyclone. Most often they are spawned in areas of strong updrafts associated with severe thunderstorms, near cold frontal boundaries. A very strong jet stream will often provide the mechanism of shearing at upper levels that starts the process of rotation. A severe thunderstorm is one that either has winds that reach or exceed 58 miles per hour or contains hail greater than or equal to ¾ inch in diameter, or that has both characteristics. Tornadoes also may form in mesoscale convection systems: slow-moving, long-lasting, nearly circular clusters of interacting thunderstorms. Tornadoes associated with such systems are often weak. These are the same conditions that lead to the formation of hail along with strong down drafts inside the thunderstorm; hail often accompanies tornadoes. These strong downdrafts can lead to the production of very strong straight-line winds, as opposed to the rotating motion of a tornado. Macrobursts (larger) and microbursts (smaller) are the two specific types of downbursts. Although the damage is usually not as extensive as in a tornado, these straight-line winds, or downbursts, can cause extensive damage. Hurricanes also can spawn tornadic activity.

Fortunately, tornado outbreaks are relatively rare in New England, compared to frequencies on the Great Plains of the United States, including Texas,

Fig. 17.1. Tornadoes in New England. Photograph of June 1953 Worcester tornado (*top*) and the second Martha's Vineyard waterspout, 1:02 P.M., 19 August 1906 (*bottom*). Top photo from Ludlum (1976; p. 47). Bottom photo from National Weather Service (National Oceanic and Atmospheric Administration/Department of Commerce) catalog of historical photos http://www.photolib. noaa.gov/historic/nws/nwind1.htm.

Oklahoma, Kansas, and Nebraska. In fact, as a geographical region, New England experiences the fewest tornadoes of any region east of the Rocky Mountains (fig. 17.2). Despite the infrequent occurrence of these violent storms, tornadoes have been documented in all corners of New England, from the Allagash Valley, Maine, in the northeast, to Nantucket Island in the southeast, and from Saint Albans, Vermont, in the northwest to Greenwich, Connecticut, in the southwest (fig. 17.3; Ludlum 1976). The average New England tornado occurs in mid- to late summer (July and August) in the late afternoon. Rarely does a tornado hit New England before May. Some are possible September through part of November. The common path traveled by most tornadoes is from southwest to northeast, but New England has many tornado tracks that move from northwest to southeast or even west to east (fig. 17.2). We believe this is related to storm tracks across New England, and particularly to cold fronts moving across the region from northwest to southeast.

Massachusetts has the highest number of documented tornadoes in New England, and the highest annual average, while Rhode Island has the smallest number of documented events (table 17.1; fig. 17.3). The area most affected by tornadoes lies just to the east of the Berkshires in north-central Massachusetts (Leathers, 1994). Maine, New Hampshire, and Connecticut are close in number in their statewide tornado occurrence rates, averaging between 1.4 and 1.8 per year for the period 1950 to 1996. These averages pale in comparison to states such as Texas and Oklahoma, which reported 5,860 (125 per year) and 2,420 (51 per year), respectively, over this same time period. Similarly, hail occurs in New England in only slightly over 1 to less than 1 day per year, on average, reflecting the low number of severe thunderstorms in the region compared to other parts of the country.

Table 17.1
Number of documented tornadoes
for each New England state, 1950–1996

State	Tornado reports 1950–1996	Average per year
Massachusetts	134	2.9
Maine	83	1.8
New Hampshire	73	1.6
Connecticut	65	1.4
Vermont	32	0.7
Rhode Island	8	0.2

Modified from Leathers (1994).

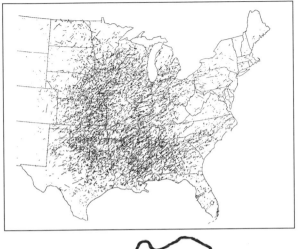

Fig. 17.2. Paths of significant tornadoes between 1880 and 1989 (*top*), and "blow-up" of those in New England (*bottom*). Arrow pointing to Massachusetts also indicates the path of the 1953 Worcester tornado. Used with permission from Tom Grazulis (1991), The Tornado Project.

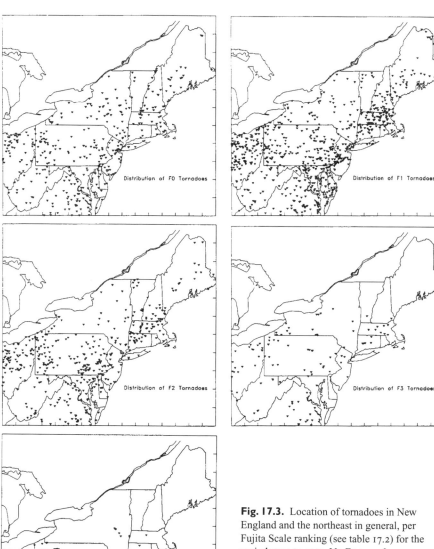

Fig. 17.3. Location of tornadoes in New England and the northeast in general, per Fujita Scale ranking (see table 17.2) for the period 1950 to 1990. No F5 tornadoes were recorded in New England during that time period. Modified from Leathers (1994; fig. 4, pp. 11–12), as available from the Northeast Regional Climate Center (National Oceanic and Atmospheric Administration/Department of Commerce), Cornell University.

New England receives far fewer tornadoes than most other states because of its relative location. It is situated far enough north that the jet stream is located south of New England for much of the year, leading to the presence of relatively cool temperatures in the region. Cooler temperatures help stabilize the atmosphere, which suppresses opportunities for strong updrafts in the atmosphere, thus inhibiting the development of tornadoes. In summer (July and August, in particular), the jet stream moves farther north, bringing warmer temperatures, more humidity, and greater instability to the atmosphere around New England. This instability results in greater thunderstorm activity and the potential for tornado development. However, the cold water off the east coast of New England in summer reduces the intensity of thunderstorms that may eventually produce tornadoes and hail. As a result, tornadoes are rare within about 15 miles of the coast. Although the values in table 17.1 are not adjusted for area, the relatively small number of events in Rhode Island is partly driven by its maritime location, even though the water temperatures are warmer in this region than along the eastern coastal zone of New England. The overall cooler climate of northern New England greatly inhibits the growth of severe storms, especially because the flow of warm moist air from the Gulf of Mexico or the sub-tropical part of the North Atlantic must travel over long distances to get to northernmost New England.

Tornado Impact

At their worst, tornadoes are unmatched when it comes to destruction. Fortunately, most tornadoes are relatively weak, with winds speeds slower than those associated with a hurricane. These weaker tornadoes (F0 on the Fujita Tornado Intensity Scale) typically snap trees and damage roofs on houses, whereas strong tornadoes (F3 or greater) leave behind near-total devastation (table 17.2 and fig. 17.4). Strong tornadoes have been reported to pick up train locomotives and toss them tens of yards away. Needless to say, mobile homes pose no challenge at all to most moderate to strong tornadoes, and injuries and fatalities can be great.

Hallmark Tornadoes

The six deadliest New England tornadoes are presented in table 17.3. By far the worst to strike in New England was the Worcester tornado of 9 June 1953. The Worcester tornado touched down at Petersham, Massachusetts, and took a path east-southeastward to Southboro, covering 46 miles and lasting an hour and twenty minutes (Ludlum, 1976). Nested within the same storm, two other tornadoes were spawned that day, one in Exeter, New Hampshire, the other in Sutton, Massachusetts. Clearly, this was one of the worst tornado days in New England history, with ninety deaths from one tornado alone (mostly in Worcester) and ninety-four from the three tornadoes combined (Ludlum, 1976). The second-worst tornado from 1950 to 1990 led to only eight deaths. Seventeen tornadoes between 1880 and 1995 are known to have killed individuals within

Table 17.2
Fujita Tornado Intensity Scale

Fujita category	Wind (knots)	Wind (mph)	General classification	Damage (example of severity)	Percentage of all tornadoes
F0	<62	<72	Weak	Light (branches broken)	69
F1	63–97	73–112	Weak	Moderate (trailer homes damaged)	
F2	98–136	113–157	Strong	Considerable (roofs torn off houses)	29
F3	137–179	158–206	Strong	Severe (cars lifted off ground)	
F4	180–226	207–260	Violent	Devastating (houses destroyed)	2
F5	227–276	261–318	Violent	Incredible (steel structures destroyed)	

Percentage of all tornadoes is by general classification (that is, weak tornadoes comprise 69 percent of all tornadoes).

Table 17.3
Most deadly New England tornadoes, 1880–1996

County, State	Date	Strength	Estimated wind speed (mph)	Injured	Killed
Worcester, Mass.	9 June 1953	F4	207–260	1,288	90
Essex, Mass.	26 July 1890	F3	158–206	63	8
Columbia, N.Y.; Berkshire, Mass.	28 Aug. 1973	F4	207–260	31	4
Hampton, N.H.	4 July 1898	F1	73–112	120	3
Hartford, Conn.; Hampden, Mass.	3 Oct. 1979	F4	207–260	500	3
Berkshire, Mass.	29 May 1995	F3	158–206	24	3

From Grazulis (1991).

Fig. 17.4. Contrasting nature of tornado damage associated with New England tornadoes. The top photograph shows an overturned plane at the Bradley Air Museum from the 3 October 1979 tornado in Windsor, Connecticut, while the bottom photograph shows building damage from the 26 July 1890, Lawrence, Massachusetts, tornado. Top photo used with permission from Yankee Publishing Inc. as available in the Yankee Archives. Original photographer Robert H. Stepanek. Bottom photo from National Weather Service (National Oceanic and Atmospheric Administration/Department of Commerce) catalog of historical photos http://www.photolib.noaa.gov/historic/nws/nwind1.htm.

New England. Note that even weaker tornadoes can take lives, as was the case on 4 July 1898 in Hampton, New Hampshire. Similarly, lives may be lost from downbursts (especially through falling trees) and lightning from severe thunderstorms that do not produce tornadoes. For example, on 18 August 1991, the day before the arrival of Hurricane Bob, straight-line winds killed three people, while injuring eleven others in Stratham, New Hampshire. At around

Fig. 17.5. Damage associated with the 9 August 1878 tornado in Wallingford, Connecticut. Photo from The Connecticut Historical Society, Hartford, Connecticut.

four o'clock in the afternoon, several people crowded under a pavilion at the park to escape the rain. About a half-hour later, severe thunderstorms launched the straight-line winds, which toppled over the pavilion onto those who were seeking shelter under it.

There are accounts of many other tornadoes back into the 1800s and 1700s (fig. 17.5). In fact, Ludlum (1970) had compiled accounts of tornadoes from 1586 to 1870 from across the nation, including a section on New England. One of the most significant events from this early part of New England history was the four-state swarm of tornadoes on 15 August 1787 (Ludlum, 1970, p. 12). This sequence of tornadoes apparently began in central Connecticut, with touchdowns recorded in Massachusetts, Rhode Island, and eventually southern

New Hampshire. The southwest to northeast swath of tornadic activity covered an area 30 miles wide by 145 miles long. Many accounts described roofs being ripped off of buildings and damaged trees. Not many critical injuries were associated with the storms, although eight individuals in Rochester, New Hampshire, were carried several miles with their house. Another great tornado was the "Great New Hamsphire Tornado" of 9 September 1821. Although the greatest damage was from just north of Claremont on the Vermont-New Hampshire border to just northwest of Concord, sightings of other tornadoes were recorded this day in Vermont and Massachusetts. This tornado was responsible for some fatalities and serious injuries, particularly in the Kearsage, New Hampshire, area.

Figure 17.6 displays an annual time series of tornado frequencies in New England. This series includes all tornadoes of F1 strength or greater on the Fujita scale, which corresponds to wind speeds of 73 miles per hour or more (table 17.2). This plot does not include F0 tornadoes, the weakest class of tornadoes. As shown, few powerful tornadoes hit New England near the turn of the twentieth century, with some clustering of events beginning near 1950 and continuing into the early 1970s. The lower numbers in the early portion of the time series could simply be the result of lower population densities, lower reporting rates, and poor communications. The year with the highest number of tornadoes on record was 1972, when four events were recorded in New England. Frequencies appear to have declined in the 1980s and 1990s, when global

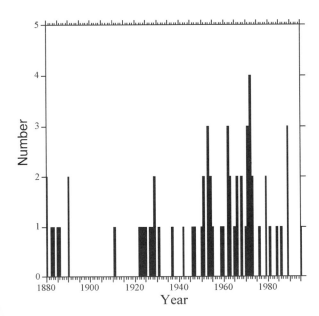

Fig. 17.6. Recorded number of tornadoes per year in New England from 1880 to 1995.

temperatures were generally above normal. Does this decline have any relation to climate change? It is probable that the recent decline is weather related and not a societal artifact, like that afflicting the earlier portion of the time series. This decline over the past two decades, however, is in contrast to the rest of the United States, where the 1990s have seen unprecedented numbers of reported tornadoes. It is unclear why New England tornadic activity may be lower recently than activity in other parts of the country affected by these deadly storms.

Summary

Tornadoes have the potential to be the most deadly storms on Earth because of the tremendous winds they produce. Luckily, New England is not affected by these storms nearly as often as the central part of the country. This does not mean that New Englanders should ignore the potential impact from these storms, as we have had our share of significant and deadly tornadoes. the majority of New England tornadoes occur in a band from southern Vermont and New Hampshire, through western Massachusetts into northern Connecticut. If there were such a thing, this is New England's "Tornado Alley." In addition, most tornadoes are not powerful in this region, although occasionally one can reach F4 status, like the 1953 Worcester tornado, the hallmark New England tornado. Most tornadoes in this region form in July and August, in the late afternoon. They often occur as an approaching cold front destablizes the very hot and humid air in place over parts of New England. Although not the most frequent type of storm that strikes New England, tornadoes can still occur and they can pack enough of a punch that individuals should heed severe thunderstorm warnings and tornado watches.

Chapter 18

Hurricanes

An immense wall of ocean rolled toward us. —AUBREY GOULD,
ON BOARD THE *BOSTONIAN*, STONINGTON, CONNECTICUT,
21 SEPTEMBER 1938

The combination of extreme wind, rainfall, waves, and the scale of a hurricane make these storms the most powerful and damaging type of storm that exists. Although born in the tropics of the Atlantic Ocean, many of these storms have the potential to wreak havoc anywhere in the mid-latitudes of the East Coast of the United States, including New England. In fact, the impact on New England can take three different forms: direct landfall in New England, landfall of a tropical system that now has extratropical characteristics (as explained later), or landfall somewhere else in the southern or eastern United States with the remnants of the storm reaching New England as a potent low pressure system (fig. 18.1). For purposes of this discussion, we consider hurricanes that cross over Long Island prior to hitting the southern cost of New England as direct landfalling hurricanes. The width of Long Island probably has minimal effect on the overall impact of the hurricane as it crosses the island. Regardless of which of the three cases exist, a hurricane that reaches New England can cause severe damage and loss of lives, despite the highly improved warning systems now in place.

In this chapter, we first give a brief introduction describing how a hurricane forms, followed by a discussion of what controls its path toward New England. Then we explore the type of damage that may be produced in New England from New England landfalling hurricanes and from those making landfall elsewhere in the country. A summary of how many hurricanes have made landfall in New England over the last century follows, with a presentation of New England's hallmark hurricanes.

Anatomy of a Hurricane

Hurricanes originate in the warm waters of the tropical oceans. In the Atlantic Ocean basin, areas of hurricane growth occur in the Gulf of Mexico and

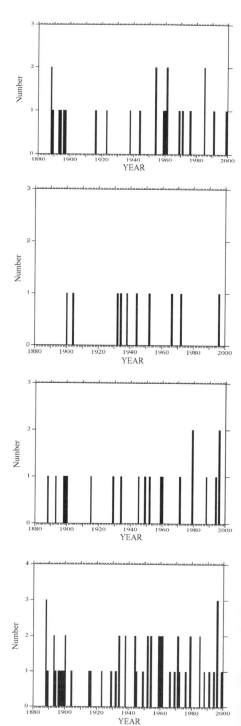

Fig. 18.1. Number of tropical systems per year from 1886 to 2000 that affected New England. The top figure shows the number of landfall hurricanes that have hit New England, while the next figure shows the number of tropical systems that made landfall in New England with extratropical characteristics. The third figure shows the number of tropical systems that affected New England although landfall occurred somewhere else in the United States. The bottom figure shows the total number of tropical systems that affected New England per year for the period 1886 to 2000.

Carribean Sea primarily in June to July and again in October to November, and in the eastern Atlantic around the Cape Verde Islands off the coast of Africa. The area of origin around the Cape Verde Islands is most active in August and September. The official hurricane season in the Atlantic basin is June through November. Hurricanes most likely to reach New England are those that originate around the Cape Verde Islands (fig. 18.2). Thus, the main season for New England hurricanes is late August through September and possibly into early October. Hurricanes can reach New England at other times during the hurricane season, but those are rare.

A hurricane originates from the development of a low pressure system or wave in the tropical Atlantic. In the case of Cape Verde hurricanes, these storms often develop out of thunderstorms that move off the coast of the Sahel region in western Africa. In fact, scientists such as Dr. W. M. Gray at Colorado State University can estimate with a reasonable amount of reliability the number of hurricanes, including the number of strong hurricanes, for the following year by evaluating present climatic conditions in the Sahel region of Africa in conjunction with other indicators. Other considerations include the state of El Niño or La Niña (see chapter 5). Under El Niño conditions, strong upper-level west to east winds in the tropics have the potential to shear off the tops of thunderstorms in tropical depressions, thereby inhibiting tropical storm and ultimately hurricane development. In contrast, the La Niña mode of the tropical Pacific Ocean creates favorable upper-air conditions for hurricane development, as well as major hurricane development. Neutral conditions in the El Niño–La Niña system will often produce an intermediate or closer to average number of hurricanes.

Once these low pressure systems form over the tropical oceans, convection

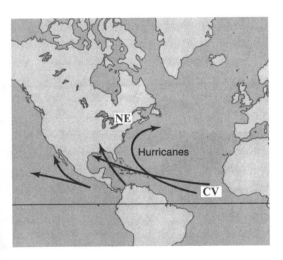

Fig. 18.2. Tracks commonly taken by hurricanes in the North Atlantic that may have an impact on New England. Most storms that make landfall in New England (*NE*) originate near the Cape Verde Islands (*CV*) off the coast of Africa. Modified from METEOROLOGY: 5/E, THE ATMOSPHERE AND THE SCIENCE OF WEATHER, by Moran/Morgan, © 1997 (fig. 15.6, p. 360). Reprinted by permission of Pearson Education, Inc., Upper Saddle River, N.J.

can increase markedly provided that the temperature of ocean surface waters is at least 80°F. Below this temperature, hurricanes will not form because not enough convection exists to fuel hurricane growth. Hurricanes grow by the energy released from the condensation of evaporated water off the ocean surface. With continued convection, a large number of thunderstorms will start to develop. This cluster of thunderstorms will begin to rotate because of Coriolis forces generated by the spin of Earth. Once the cluster takes the form of a well-developed circular storm, convection will continue to decrease the central pressure of the swirling mass, creating a low pressure system, or tropical depression. Air converging inward toward the center of the spiral further increases convection and wind speeds. Once sustained surface winds (not gusts) reach a maximum of 39 miles per hour, tropical storm status is achieved. Hurricanes form when maximum sustained winds reach 74 miles per hour, the time when an eye and distinct eye wall may form. Once becoming a hurricane, the relative strength and damage potential for an individual storm is characterized through the Saffir-Simpson intensity scale. Table 18.1 shows this rating system. Note that tropical storms also may reach New England with the potential for devastating effects.

Once a hurricane has formed in the central Atlantic, such as a Cape Verde hurricane, it will travel to the west on the tropical easterlies. As it approaches the western part of the Atlantic and the United States, it often will drift northward as it starts to come under the effect of the westerlies. Hurricanes will then travel in a northerly and eventually northeasterly direction as they move around the Bermuda-Azores High (fig. 18.2). The position and strength of this high, plus the strength and pattern of the westerlies, as well as that of any mid-latitude cyclones and anticyclones in the region, will determine the path that the hurricane takes once it moves closer to the United States. Weaker wester-

Table 18.1

Saffir-Simpson Hurricane Intensity Scale

Tropical System	Category	Pressure (mb)	Pressure (inches Hg)	Sustained winds (knots)	Sustained winds (mph)	Storm surge (ft
Tropical depression	N/A	N/A	N/A	<34	<39	N/A
Tropical storm	N/A	N/A	N/A	34–63	39–73	N/A
Hurricane	1	>980	>28.94	64–82	74–95	4–5
Hurricane	2	980–966	28.91–28.58	83–95	96–110	6–8
Hurricane	3	965–945	28.47–27.91	96–113	111–130	9–12
Hurricane	4	944–920	27.88–27.17	114–135	131–155	13–18
Hurricane	5	<920	<27.17	>135	>155	>18

lies and the absence of any strong frontal boundaries coming off the United States could either allow the hurricane to travel directly into the Gulf of Mexico or the southeastern United States or lead to a more erratic and unpredictible path. If the westerlies are a little stronger, the hurricane may move in a more northerly direction up the Atlantic coast toward New England. The nature of high pressure systems and fronts in the northeast will then influence whether the hurricane can take direct aim on New England or whether it will be pushed out to sea before reaching New England. Most direct hits in New England are on the southern coast of Connecticut and Rhode Island or Cape Cod. Some may move beyond Cape Cod and strike the southern Maine coast. It would take a very unique set of circumstances for a hurricane to make initial landfall around Boston or in the Casco Bay area of Maine.

As a hurricane moves northward toward the mid-Atlantic states and New England, it will weaken because of the cooler ocean surface temperatures. These cool waters mean less convection and less fuel for the hurricane to maintain its strength. It is for that reason that most hurricanes reaching New England will be, at best, a category 3 storm by the Saffir-Simpson intensity scale. However, a category 3 hurricane is considered a major hurricane, and damage can be excessive. Sometimes a hurricane will start to take on extratropical storm characteristics before landfall, as colder air starts to move into the core of the storm. At this stage, the satellite view of the hurricane will start to become more comma-like, typical of mid-latitude, extratropical storms (fig. 14.6). Likewise, hurricanes or even tropical storms that make landfall south of New England may eventually work their way up the East Coast much like a mid-latitude low pressure system. However, these storms can still carry copious amounts of tropical moisture. Furthermore, as was discussed in chapter 14, hurricanes in the Atlantic basin may interact with mid-latitude systems to produce tremendous rainfall amounts, as in the 1996 October rains in eastern New England, or powerful storms that can wreak havoc on maritime communities, as was the case for the 1991 "All Hallows Eve" storm (see chapter 15).

Hurricane Impact

Damage and the resulting costs associated with hurricanes making landfall in New England primarily come from flooding with the heavy rainfall and possibly from the storm surge, depending on where landfall was attained (table 18.2 and fig. 18.3). It is noteworthy that of the twenty-five most-costly hurricanes to hit the United States, six had a major impact on New England. Lesser damage probably comes from the wind, simply because most hurricanes reaching New England do not have the 131-plus mile per hour winds of a category 4 storm (table 18.1). The strongest storms are minimal category 3 storms with maximum

Fig. 18.3. Flooding in Ware, Massachusetts, associated with the Hurricane of '38. Note the remains of the stone bridge (left-center) that was destroyed by the flooding. Photograph from the National Weather Service (National Oceanic and Atmospheric Administration/Department of Commerce) catalog of historical photos at http://www.photolib.noaa.gov/historic/nws/nwind1.htm.

sustained winds of 111 to 130 miles per hour. Historically, most deaths from hurricanes moving into New England probably come from the storm surge, as was the case for the 1938 hurricane.

Of all the potential causes of property damage in a New England hurricane, rain and its associated flooding probably has the largest impact. Hurricanes and tropical storms as far north as New England can still produce 10 inches or more of rain. In the case of Hurricane Diane in 1954, which passed just south of New England, over 18 inches of rain fell in north-central Connecticut and south-central Massachusetts (see tables 14.1 and 18.2). Excessive amounts of rain are also the means by which hurricanes making landfall somewhere else in the contiguous United States can affect New England. Two striking examples of this latter scenario occurred in the late 1990s. Hurricane Bertha came ashore close to the South Carolina–North Carolina border, but it was able to produce 4 to 5 inches of rain over a large area of New England. The highest one-day rain total associated with Bertha was 5.22 inches at West Rockport, Maine, on 14 July 1996. Three years later, Hurricane Floyd also came ashore in the Carolinas, but it produced 4 to 6 inches of rain over wide areas of New England. The one-day total of 9.92 inches on 17 September 1999 from Floyd at Mount Mansfield, Vermont, is one of the highest one-day storm totals in New England (table 14.1).

Storm surge is the wall of water in the northeast quadrant of an approaching hurricane that is pushed ahead by the forward speed of the hurricane and the strong south to southeast winds on the right side of the hurricane (fig. 18.4). The very low pressure around the eye of the hurricane also causes the water at the center to "bulge," increasing water levels even more. Water will rise about 1 foot every 0.89 inches of mercury (30 millibars) decrease in surface pressure. Thus, a hurricane with a central pressure of 950 millibars will have a 1-foot higher storm surge due to this pressure bulge than a hurricane with a central pressure of 980 millibars. As the hurricane makes landfall, this wall of water from the combined wind and pressure components of the storm will be pushed ashore, flooding many low-lying areas and washing away structures along the coast. Damage is even greater when landfall occurs at the same time as high tide. Storm surge was one of the main contributors to the high death toll from the 1938 New England hurricane that made landfall across Long Island at the Connecticut–Rhode Island border. The impact of the storm surge was compounded by the shape of the coastline, particularly the narrow embayment of

Table 18.2
Damage for twenty-five most costly hurricanes
to make landfall in the United States

Rank	Hurricane name (State/Area)	Year	Category	Damage (1996 dollars)
1	Andrew (Fla./La.)	1992	5	30,475,000,000
2	Hugo (S.C.)	1989	4	8,491,581,181
3	Agnes (northeast U.S.)	1972	1	7,500,000,000
4	Betsy (Fla./La.)	1965	3	7,425,340,909
5	Camille (Miss./Ala.)	1969	5	6,096,287,313
6	**Diane (northeast U.S.)**	**1955**	**1**	**4,830,580,808**
7	Frederic (Ala./Miss.)	1979	3	4,328,968,903
8	**New England**	**1938**	**3**	**4,140,000,000**
9	Fran (N.C.)	1996	3	3,200,000,000
10	Opal (Fla./Ala.)	1995	3	3,069,395,018
11	Alicia (Tex.)	1983	3	2,983,138,781
12	**Carol (northeast U.S.)**	**1954**	**3**	**2,732,731,959**
13	Carla (Tex.)	1961	4	2,223,696,682
14	Juan (La.)	1985	1	2,108,801,956
15	Donna (east U.S.)	1960	4	2,099,292,453
16	Celia (Tex.)	1970	3	1,834,330,986
17	Elena (Gulf Coast)	1985	3	1,757,334,963
18	**Bob (N.C./northeast U.S.)**	**1991**	**2**	**1,747,720,365**
19	Hazel (S.C./N.C.)	1954	4	1,665,721,649
20	Florida (Miami)	1926	4	1,515,294,118
21	Texas (Galveston)	1915	4	1,346,341,463
22	Dora (Fla.)	1964	2	1,343,457,944
23	Eloise (Fla.)	1975	3	1,298,387,097
24	**Gloria (east U.S.)**	**1985**	**3**	**1,265,281,174**
25	**Northeast U.S.**	**1944**	**3**	**1,064,814,815**

Hurricanes in bold had a major impact on New England. From the National Hurricane Center (National Oceanic and Atmospheric Administration/Department of Commerce) web site at http://www.nhc.noaa.gov/.

Fig. 18.4. Storm surge in Connecticut associated with Hurricane Carol, 31 August 1954 (*top*). Damage is increased when high waves come onshore on top of the storm surge, as occurred during Hurricane Carol (*bottom*). Photographs from the National Weather Service (National Oceanic and Atmospheric Administration/ Department of Commerce) catalog of historical photos at http://www.photolib.noaa. gov/historic/nws/nwind1.htm.

Narragansett Bay. Although storm surge is more commonly associated with hurricanes, the 2 February 1976 nor'easter produced a storm surge up the Penobscot embayment of Maine, as we described in chapter 15.

Wind damage to structures is probably not that great along the coast because of the moderate strength of hurricanes reaching New England, but further inland, damage related to the uprooting of trees may be extensive. For example, the 1938 hurricane was responsible for tremendous tree loss from wind. It is estimated that the hurricane uprooted some 275 million trees, resulting in about 2.6 billion board feet being thrown down (Blue Hill Observatory, 1998). Much of this damage was in Massachusetts, New Hampshire, and Vermont. With tree fall, power outages are a major concern. The likelihood of major damage is increased as the soil becomes saturated with the heavy rains brought on by the storm system. Slower-moving storms and the remnants of hurricanes can still have wind gusts in the 30 to 40 mile per hour range, more than enough to uproot even the largest trees in saturated soil. Records of the tracks taken by past hurricanes began in 1886. This 100-plus-year record gives us the opportunity to evaluate how hurricanes and tropical storms have made landfall in New England, as well as how many hurricanes with extratropical characteristics made landfall in New England and the number of hurricanes that made landfall elsewhere, but still affected New England. We analyzed all tropical systems whose track took them through New England during the time they were followed by the National Hurricane Center. We also identify periods when the total number of tropical systems had an impact on New England. A total of fifty-one tropical systems have had an impact on New England over the past 110 years, for an average of once every 2.1 years (fig. 18.1). During that time, two systems impacted New England in the same year in fifteen years (once every 7.7 years), and three systems affected the region twice, in 1888 and 1996 (once every 57.5 years). At least one hurricane made landfall in New England in twenty years, resulting in an average of once every 4.8 years. On the other hand, two hurricanes made landfall in New England four times—in 1888, 1954, 1961, and 1985—or once every 28.8 years. It is interesting that the hallmark blizzard also occurred in 1888, which makes one ponder about how New Englanders perceived this "increase in extreme events." We can only wonder if New Englanders believed that they were being hit with more extreme weather events than in previous years, much the way many individuals wondered if we saw more extreme weather events over the decade of the 1990s than in previous decades.

As for the other types of tropical systems, at least one hurricane with extratropical characteristics made landfall in New England in ten years, or once every 11.5 years. Two such storms have not struck in one year anytime since these records began in 1886. Eighteen hurricanes made landfall somewhere else over the last 115 years that had an impact on New England. This is an average of about once every 5.8 years. In two years—1979 and 1996—two hurricanes making landfall elsewhere had an impact on the region.

Given that thirty-seven years over the last 115 years have seen at least one tropical system affecting New England, the average period of time between years with at least one tropical system is about three years, so we have a 33 percent chance that a tropical system will hit New England in any given year. The fifteen years with at least two tropical systems over the last 115 years result in an average of one year in about eight having two tropical systems or about a 12 percent chance of having a year with two tropical systems. Three systems struck in a single year 2 in 115 years, for a 2 percent chance of having three tropical systems affect New England in the same year.

Over the last 115 years, there are two time periods when the number of tropical systems affecting New England was the highest. During the seventeen-year period from 1888 to 1904, inclusive, fourteen systems affected New England, similar numbers to the fifteen tropical systems that affected New England over the twenty-one-year period from 1952 to 1972 (fig. 18.1). In addition, there were five years with at least two tropical systems affecting the region during that latter time frame, surely a very active period for New England hurricanes. On the other hand, several time periods show tropical systems rarely having a role in the New England climate. Over the twenty-eight years from 1905 to 1932, inclusive, there were only five systems, an average of only 1 every 5.6 years, just about half of the 115-year average.

Fig. 18.5. An example of the terrible amount of destruction along the Rhode Island coast associated with the storm surge and waves from the Hurricane of '38 is shown by this sequence of photographs of Watch Hill, Rhode Island, before (*top*) and after (*bottom*) the hurricane came ashore. The two groins in the center of the bottom photo are seen in the left-center portion of the top photo. Photographs from Watson (1990; pp. 6–7). Used with permission from David Lucey, *The Westerly Sun.*

Fig. 18.6. Waves crashing along a seawall in Rhode Island during the Hurricane of 1938. Photograph from the National Weather Service (National Oceanic and Atmospheric Administration/Department of Commerce) catalog of historical photos at http://www.photolib.noaa.gov/historic/nws/nwind1.htm.

The periods with the greatest number of landfall hurricanes are similar to those for all tropical systems, with a cluster between 1888 and 1897, that is, seven landfall hurricanes over a ten-year period. The period from 1954 to 1961 is similar with six landfall hurricanes over an eight-year period. The most quiescent period goes from 1898 to 1937, when only two landfall hurricanes occurred or two over a thirty-nine-year period. However, the following year, the New England hurricane of all time struck the region.

Hallmark Hurricanes

Without a doubt, the 21 September 1938 hurricane is the standard by which all other New England hurricanes are measured. Hurricanes were not labelled until 1950, with the use of people's names originating in 1953. The 1938 storm was responsible for over six hundred deaths and $3.4 billion in damage (in 1998 dollars) in New England (also see table 18.2). Most lives lost were due to the storm surge and waves along the Rhode Island coast, and particularly the storm surge up Narragansett Bay (figs. 18.5, 18.6, and 18.7). However, loss of life from a similar hurricane today probably would be lessened because of satellite and other advanced warning systems. The curious fact about the 1938 hurricane is that the National Weather Service (NWS) failed to realize how strong the storm actually was. This is despite the warnings of a junior meteorologist at the NWS and the exceptionally low barometer readings from a ship close to Florida at the time the storm was in that area of the ocean.

Fig. 18.7. Classic photo of the storm surge that flooded downtown Providence, Rhode Island, during the 1938 Hurricane. Photograph from Watson (1990; p. 8). Photographer unknown.

Fig. 18.8. Flooding along the Merrimack River in Manchester, New Hampshire (*top*), and along the Connecticut River in Hartford, Connecticut (*bottom*), during the 1938 Hurricane. Photographs from the National Weather Service (National Oceanic and Atmospheric Administration/Department of Commerce) catalog of historical photos at http://www.photolib.noaa.gov/historic/nws/nwind1.htm.

On the other hand, property losses could easily be greater today because there is more development along the coastline. It is also easy to forget the great loss of trees along the path of the hurricane as it moved inland through Vermont and New Hampshire with a forward velocity of over 50 miles per hour. At the same time, extensive flooding occurred, especially in western New England (fig. 18.8). Flood levels on the Connecticut River from the 1938 hurricane have been exceeded only by the 1936 "All-New England Flood," produced by heavy rains on top of snow.

After the 1938 storm, the strongest to hit New England was Hurricane Gloria in 1985. It had a central low pressure of 951 millibars with maximum sustained winds of 85 knots per hour (a minimum category 3 hurricane) when it made landfall in western Long Island. It had reached category 4 at its peak. Gloria quickly moved into Connecticut, then up the Connecticut River valley

Fig. 18.9. Storm surge in Buttermilk Bay, Buzzards Bay, Massachusetts (*top*), from Hurricane Bob. Although most damage to coastal regions occurred in Rhode Island and Massachusetts, the storm surge and waves also had an impact along the New Hampshire and Maine coasts, as evidenced by this boat left high-and-dry in Camden, Maine. Photographs from Minsinger and Orloff (1992; pp. 56 and 76). Top photo used with permission from Photographer—from Hurricane Bob—Blue Hill Observatory, 1991. Bottom photo © 1991 Bangor Daily News, used with permission.

of western Massachusetts. Although the storm lost its punch very quickly, it did produce a significant amount of damage (table 18.2).

The most recent hurricane to inflict significant damage in New England was Hurricane Bob in 1991 (table 18.2). Bob came ashore in Rhode Island on 19 August, with the eye moving over Newport. Bob was only a category 1 storm by the time it arrived in New England, with a central low pressure of 964 millibars and maximum sustained winds of 85 miles per hour. Peak gusts from Bob were 91 knots (105 mph) on Block Island, while up to 6 inches of rain fell in many places in eastern New England. The storm surge east of Newport inflicted heavy damage on many areas, particularly from boats being pushed ashore (fig. 18.9). Seventeen lives were lost in this storm, and it caused $1.5 billion in damages by the time the clean up was completed.

The closest-spaced hurricanes, in time, to hit New England were Carol (31 August) and Edna (11 September) in 1954. Carol was a category 2 storm when it hit, with perhaps the most famous impact of this storm being the toppling of the steeple of the Old North Church, Boston. However, Carol was responsible for the loss of fifty lives and significant monetary losses. Although neither of the two August 1955 hurricanes Connie or Diane made landfall in New England, their impact may have been more influential than many landfalling hurricanes that hit the region. As we discussed in chapter 14, the heavy rains, particularly the record-setting rains from Diane, caused extensive flooding and property damage in western Massachusetts (see tables 14.1 and 18.2).

Prior to the tracking of hurricanes in the 1880s, the most famous hurricane is the Great September Gale of 1815 (Ludlum, 1963; fig. 18.10). Considered

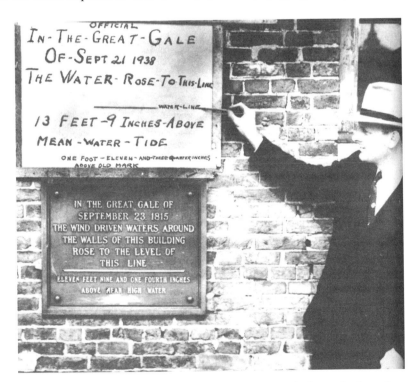

Fig. 18.10. Storm surge for the September Gale of 1815 was only about 2 feet lower than that for the Hurricane of 1938, as evidenced by the plaques set on Providence's Old Market House. Photograph from Chales S. Williams, Jr./*The Providence Journal* as seen in B. Watson, *New England's Disastrous Weather* (1990, p. 12).

the second most powerful storm with an equally great impact, this storm has been long considered the hallmark New England hurricane behind the hurricane of 1938. The storm appears to have come across eastern Long Island on 23 September, close to where the 1938 hurricane came onshore. The 1815 gale probably moved into Connecticut around New Haven. It appears to have moved through Connecticut and central Massachusetts before pushing into southwestern New Hampshire. Damage associated with the storm came from the high tides caused by the storm surge, wind damage to many structures, and heavy rains (fig. 18.10). New Haven had over 6 inches of rain on 22 and 23 September.

Another hallmark hurricane that struck during the early period of colonization was the Great Colonial Hurricane of August 1635 (Ludlum, 1963). The notoriety of this storm came from the accounts of John Winthrop of the Massachusetts Bay Company and William Bradford of Plymouth Plantation. Complete forests were leveled by the wind, indicating that it was still a very powerful storm when it reached New England. Other evidence of the storm's ferocity included the driving of ships from their anchors and overthrowing of houses. The storm seemed to be the worst in the Cape Cod and Narragansett Bay areas of New England.

Summary

Hurricanes are the largest storms on Earth. New England has felt the wrath of these behemoths in the past and will most assuredly feel them in the future. Fed by warm ocean waters in the southeastern part of the north Atlantic Ocean, these storms eventually strengthen and make their way to the eastern U.S. coast. Once there, if no strong circulation systems steer them away or take them into the southeastern part of the United States, they may move up the coast and make landfall in New England. However, hurricanes making landfall in other parts of the country can still bring large amounts of rain into New England and cause tremendous damage from the flooding. The one-day record rainfall for New England came from Hurricane Diane in 1955. The famous Hurricane of 1938 is the hallmark hurricane for the region, as its fast movement, high storm surge, and tidal flooding caused massive damage along the Connecticut and Rhode Island coastlines. Further inland, heavy rains caused extensive flooding and heavy winds toppled many trees, particularly in Vermont and New Hampshire. Although the Great September Gale of 1815 did not appear to be quite as powerful as the 1938 storm, it was close.

The tropical systems that affect New England include direct landfall hurricanes, those that take on extratropical characteristics at the time they make landfall, and those that make landfall elsewhere, but track through New England or are just off the coast. New England is affected by a tropical system, on average, once every two years. The region is hit by two events in any one year, on average, once every eight years. As far as landfall hurricanes go, New England has a landfall hurricane once every five years and two landfall hurricanes in the same year once every twenty-nine years. There have been two periods over the last 115 years when tropical systems were very prominent in the New England climatic picture, 1888 to 1904 and 1952 to 1972. We may be moving into another period of frequent hurricanes in the Atlantic Basin, but that does not necessarily transfer directly into more hurricanes affecting New England.

Part VI

Changes over Time

I have been trying merely to do honor to the New England weather—
no language could do it justice. —MARK TWAIN

In the preceding five sections of this book, we presented the many controls on New England's climate at different time scales, the spatial variability across the region, the types of weather and climate felt by New Englanders through the year via a look at individual seasons, and the extreme events that affect the area. We will now focus on the specific changes in climate that occur over longer time periods. This section will look at recent trends in temperature and precipitation across the region over the length of time available in records from individual sites. We will do this by evaluating changes within the individual states. This part of the book will emphasize the climatology of New England. Chapter 19 will look into the past through two techniques: changes over the last approximately one hundred years documented by instrumental records, and detailed records of past weather conditions available in historical writings. Chapter 20 will anticipate what may be in store for us in the years to come.

We present these views into the past and future because we feel it is important to gain perspective on how the complex New England climate system has varied and what conditions we may see in the future. If we only look at how weather systems have changed over a few years or few tens of years, our perspective is very limited and narrow. Many extreme situations can occur only once every hundred years (such as the rains of October 1996; chapter 14) or even once every millennium. A single extreme event could easily happen within one's lifetime despite an event of such magnitude never having occurred in the more recent past. It is necessary to look at records of weather and climate as long and as detailed as possible to find true insight into how New England's climate operates. A more thorough understanding and awareness of cycles and trends in particular aspects of our weather and climate allows us to make much more educated predictions about what could happen in the future. Many estimates of specific events such as one-hundred-year floods are statistically based. When the records are short, the error in predictions may be quite large. Longer records provide a much larger data set to evaluate, reducing the error and allowing meteorologists to come closer to predicting what may happen in the future. Lengthy records from the past may enable us to extend cycles or trends into the future as a predictive tool, as well as to define the magnitude of specific events within New England's climate system. We present this part of the book to provide that perspective on New England's weather and climate. This look into a continuous time series of changes in New England was not a part of Ludlum's 1976 original survey of the region's weather.

Chapter 19

Looking into the Past

*It is a popular opinion that the temperature of the winter season, in
modern latitudes, has suffered a major change, and become warmer
in modern, than it was in ancient times.* —NOAH WEBSTER, 1799

W e divide this chapter into two sections for our discussion. The first
part will be a look at how temperature has varied over the length of
the instrumental record. Data archived at the National Climate Data
Center in Asheville, North Carolina, for the two-to-three climatic regions per
New England state over the last century (except Rhode Island is a single region
as per figure 3.2), are used to show how the region's climate has changed. To
obtain a single state-wide average, the climatic zones within each state are
combined after accouning for the percentage of the area covered by each zone
within the state. These official instrumental records extend back to 1895, thus
we have a little over a century to quantify how New England's climate may
be changing. Several unofficial records extend well beyond the last century,
such as the Ezra Stiles record from New Haven, Connecticut, and the Edward
Holyoke record from Salem, Massachusetts, each of which goes back into the
late 1700s, and a record from Dartmouth College, Hanover, New Hampshire,
which goes back to 1834. That these are "unofficial" does not mean that these
records are not of use; it means that they were collected under different techniques than most of those of the last century, thus one has to evaluate these
older records more carefully. Nevertheless, one hundred-plus years is not a
very long record to understand completely how the region's climate varies, and
especially, for determining the complete range of variability in temperature
and precipitation felt by New Englanders. However, it is what we have to work
with, and the records we discuss below shed light on various aspects of New
England's climate, not only with time, but across the region. In fact, quanti-
fying the variability in climate over this past century is a key component in
deciphering what could happen in the future, not only for New England, as we
discuss in chapter 20, but for the country as a whole.

As part of a national program to evaluate how the country's climate has changed in the past and what causes those changes, regional studies recently were done around the country. The New England Regional Assessment Group produced a summary in 2001 of New England's climate over the last 107 years; their findings provide an excellent starting point to discuss past changes. We present their findings on changes in mean annual temperature and in mean summer and winter temperatures through the past century. As we discuss the long-term trends in these data, such as any evidence that reflects the accumulation of greenhouse gases, we also point out the specific years when we can see the impact of past El Niño events, La Niña events, and volcanic eruptions.

The second part of this section is a discussion of the continuous record of precipitation, both total precipitation (liquid equivalent) and snowfall amounts, over the length of the instrumental record. We again will suggest how various climatic forcing factors may be observed in the precipitation record, although the erratic nature of precipitation events compared to temperature distribution at a single site makes interpretation of past precipitation trends more complicated. We also evaluate the spatial variability throughout New England, particularly how climate has varied inland compared to the coast.

The final section in this chapter will highlight and explain through specific examples how we can extend the climatic record on a daily basis back into the 1600s, through the entire 1700s, and up to the time when instrumental records became prevalent in the late 1800s. This is accomplished through the use of the written historical record. Many New Englanders kept meticulous diaries noting accurate daily weather information. These data included thermometer readings (once they were available to the general public), qualitative temperature data (such as hot, warm, cool, cold), wind directions, precipitation type (including quantitative measurements of snow and sometimes rain), and barometer readings when that instrument became available to the general public. Many individuals also recorded dates of harvest, fruit tree blooms, and other phenological information. We will give several specific examples of how such information, as well as other written information such as annals, logs, and newspapers, can be used to reconstruct past weather in extraordinary detail. Ludlum (1976) was one of the first individuals to use the written record in this manner when he reconstructed weather conditions for specific periods during the Revolutionary War. The final part of this chapter will highlight several other specific periods when the weather played an important role within some aspect of New England society.

Instrumental Records of Past Temperature Trends

A quick and easy summary of how mean annual temperatures have changed by state over the last century—and we emphasize that this is a quick and easy

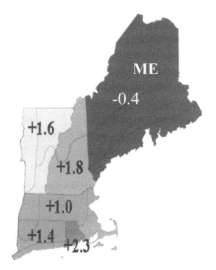

Fig. 19.1. Suggested change in mean annual temperature for each New England state between 1895 and 2000 as modified from the New England Regional Assessment Group. See figure 19.2, which shows how these numbers were derived and the changes in mean annual temperature for each state. The climatic division data from the National Climatic Data Center (National Oceanic and Atmospheric Administration/Department of Commerce) were used in the production of this figure. Climatic divisions shown by the light lines within each state and on figure 3.2. New England Regional Assessment Group. 2001. *Preparing for a Changing Climate: The Potential Consequences of Climate Variability and Change.* New England Regional Overview. U.S. Global Change Research Program, 96 pp., University of New Hampshire.

conclusion—is to say that every state in New England except for Maine has warmed by over 1°F (fig. 19.1). Maine shows a decrease in mean annual temperature of 0.4°F. Greatest warming occurs in Rhode Island, with a 2.3°F increase, followed by Vermont and New Hampshire with increases over 1.5°F. Figure 19.2 shows the actual time-series of mean annual temperature for each state from 1895 to 2000, and indicates why the amount of change in each state shown in figure 19.1 is a quick and easy number.

In 2001, the New England Assessment Group simply applied a linear fit to the data, that is, they took the temperature change where a best-fit line intersected 2000 and 1895. Although there has been much variability in temperature within each state over the last century, this simple technique does show a general increase in mean annual temperature in five of the six New England states. However, as shown by the r- and r²-values on each figure and explained in the caption, these straight lines only explain a very small portion of the variability in temperature for each state. Interestingly, the trend in temperature that most closely resembles a linear change over the last 107 years is that for Rhode Island. It is also quite interesting that the coastal zone of Maine shows about a 1°F increase in temperature since 1895 (using this same technique), the central climatic zone shows very little change, and the northern climatic zone shows a decrease in temperature of about 1°F. Perhaps the larger area covered by the northern climatic zone produces a state-wide average lowering of temperature. Clearly, there has been some control on Maine's temperature that has not affected the other New England states.

Given the great variability that characterizes New England's weather and climate, it is not surprising that there are some interesting findings in summarizing the trends in mean annual temperature for each state and for the region as a whole over the last century. The earliest parts of the record for each state were generally cold except for the late 1800s. The most exceptional example of the cool climate in the early 1900s occurred in Vermont, where even the warmer years were cooler, for the most part, than the coldest years from the 1950s to the present. The cool conditions of the earliest part of the twentieth century across the region changed rather dramatically in the late 1920s. Annual temperatures in every state except Maine increased in 1927 or 1928, remaining at high levels until 1957 to 1958. Only 1940 was a cold year during the middle

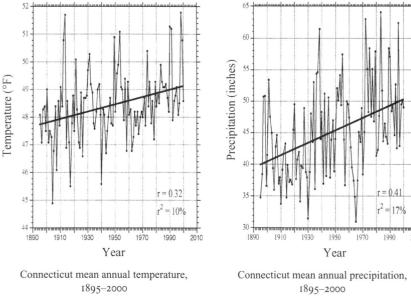

Connecticut mean annual temperature, 1895–2000

Connecticut mean annual precipitation, 1895–2000

Fig. 19.2. (*pages 234–237*) Variability in mean annual temperature (MAT) and mean annual precipitation (MAP) for each state in New England for the period 1895 to 2000. Average values for each state are achieved by areally averaging the individual climate divisions in each state. The New England states each have three climate divisions, except that New Hampshire has two and Rhode Island has one, as shown in figures 3.2 and 19.1. The solid line shows the linear best-fit for each plot. The difference between the temperature or precipitation value of the line at 1895 and at 2000 was used to suggest the amount of change in each parameter over the last 107 years, as shown in figures 19.1 and 19.4. The r-value is a function of how well the best-fit line represents the variability in the data set, with 0.99 being the highest value, and thus almost a perfect fit and indicating a linear trend to the data, and 0.01 being the lowest, and thus suggesting no trend to the data. The r^2 value (expressed as a %) indicates how much of the variability in the record can be explained by this linear fit. In almost all cases, the linear trend represents very little of the year-to-year variability in the records. Figures for each state were produced from the climate division data available from the National Climatic Data Center (National Oceanic and Atmospheric Administration/Department of Commerce), Asheville, North Carolina.

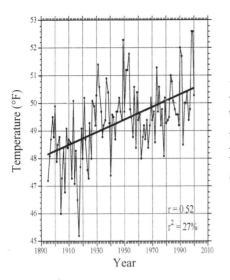

Rhode Island mean annual temperature,
1895–2000

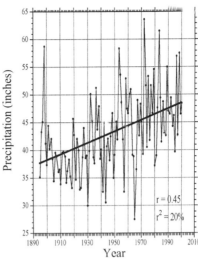

Rhode Island mean annual precipitation,
1895–2000

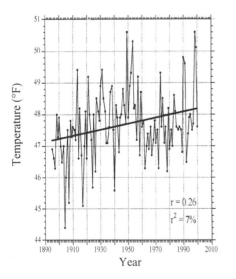

Massachusetts mean annual temperature,
1895–2000

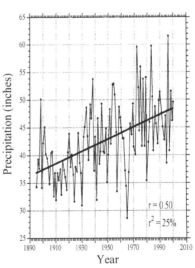

Massachusetts mean annual precipitation,
1895–2000

Fig. 19.2 *(continued)*

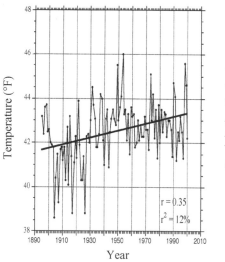

Vermont mean annual temperature,
1895–2000

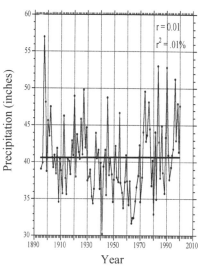

Vermont mean annual precipitation,
1895–2000

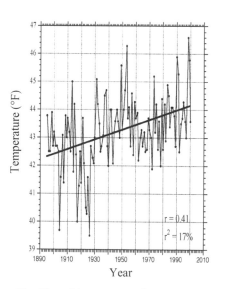

New Hampshire mean annual temperature,
1895–2000

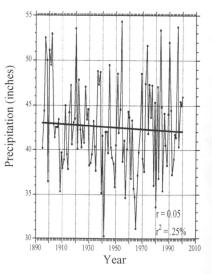

New Hampshire mean annual precipitation,
1895–2000

Fig. 19.2 *(continued)*

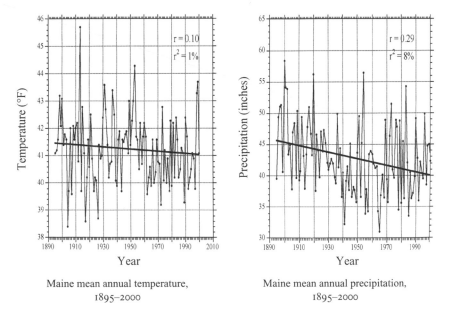

Maine mean annual temperature,
1895–2000

Maine mean annual precipitation,
1895–2000

Fig. 19.2 *(continued)*

part of this past century (table 19.1). The mid-century warming did not begin in
Maine until about 1944 and only lasted until about 1955. This decade was the
warmest overall of this past century; the late 1800s, on average, appear to have
been a little warmer. Given the difference in the long-term trend through the
century and especially in this period during the middle of the 1900s, Maine,
as a whole, truly appears to march to a different "climatic drummer" than the
other New England states. The 1960s were the coolest overall in each state, so
individuals now in their forties and fifties can tell their children that, yes, it was
much colder when we were growing up, and—as we show in the next section
on precipitation—it was snowier, too! Annual temperatures generally fluctu-
ated through 2000. However, most states have experienced some of their
warmest years on record in recent years, such as 1998 and 1999 (table 19.1).

The 2001 New England Regional Assessment Group also evaluated changes
in seasonal temperature by focusing on the extreme seasons, that is, summer
and winter. Using the same technique that they used to quantify the change in
annual temperature since 1895, they found that New England winters and summers appear to be getting warmer (fig. 19.3). Once again, the outlier is Maine,
which essentially has not seen any change in seasonal temperatures over the
century. Summer temperatures over the rest of New England have risen by
about 1°F in general, except for Rhode Island, where it appears that summer

Table 19.1

Top five hottest and coldest years and top five wettest
and driest years for each state, 1895–2001

	Temperature		Precipitation	
	Hottest	Coldest	Wettest	Driest
Maine	1913	**1904**	1900	2001
	1953	**1917**	1954	**1965**
	1999	1926	1920	1941
	1931	1972	1983	1985
	1937	1989	1901	1955
New Hampshire	1998	1926	1954	1941
	1953	**1904**	1996	**1965**
	1990	**1917**	1920	2001
	1999	1924	1983	1980
	1949	1923	1902	1957
Vermont	1953	**1904**	1897	1941
	1998	1926	1983	1963
	1949	**1917**	1990	**1965**
	1973	1907	1996	1964
	1990	1914	1927	2001
Massachusetts	1949	**1904**	1996	**1965**
	1998	**1917**	1983	1930
	1953	1907	1972	1924
	1999	1940	1975	1910
	1990	1924	1979	1941
Rhode Island	1998	**1917**	1972	**1965**
	1999	**1904**	1983	1930
	1949	1916	1898	1943
	1990	1907	1953	1957
	2001	1914	1998	1941
Connecticut	1998	**1904**	1983	**1965**
	1913	**1917**	1972	1930
	1990	1940	1996	1910
	1991	1907	1938	1924
	1953	1926	1989	1914

Years that appear in the top five for every state are in bold. One of
the top five hottest years in every state was 1953 until 2001 replaced it
as the fifth hottest in Rhode Island.

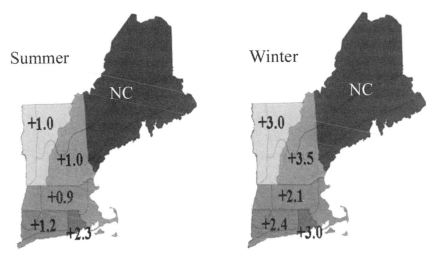

Fig. 19.3. Suggested change in mean summer (*left*) and mean winter (*right*) temperature for each New England state between 1895 and 2000. These values were determined in the same manner as those in figures 19.1 and 19.4, except that mean summer and mean winter values were plotted for the period 1895 to 2000 (similar to fig. 19.2). Modified from the New England Regional Assessment Group. 2001. *Preparing for a Changing Climate: The Potential Consequences of Climate Variability and Change*. New England Regional Overview. U.S. Global Change Research Program, 96 pp., University of New Hampshire.

temperatures may have increased by over 2°F, on average. Winter temperatures have increased by 2 to 3°F in general, with New Hampshire showing the greatest increase of 3.5°F. These numbers suggest that the transition seasons of spring and fall may actually be cooling in order for annual temperatures to have increased by only about 1 to 2°F in each state except Maine. Again, we emphasize that the technique with which these numbers were derived is quick and simple.

Nevertheless, we can identify some reasons why winter temperatures might be warming, at least in general. One possibility is a greater frequency of El Niño events in the latter part of the century. As we pointed out in chapter 3, the main impact of an El Niño event on New England is increased winter temperatures. The second possibility is less snow cover, particularly during the 1980s and much of the 1990s. Greater snow cover in the winter helps lower nighttime temperatures (see our discussion in chapters 7 and 13). No matter what the actual cause of greater warming in winter, we still wonder why Maine is not showing such a response and why the transition seasons would be colder. Does this mean winters are getting longer (starting earlier and going later), but they are not as cold? It will be interesting to see if these questions can be answered with continued research into the intricacies of New England climate.

We do see evidence in these temperature records of the specific annual to decadal climate-forcing factors that we discussed in chapter 3. Foremost is evidence of both a warming factor—that is, El Niño—and a cooling factor—that is, volcanic eruptions. One of the most distinct pieces of evidence of warming associated with an El Niño event is the warmth of 1998. This is one of the warmest years in most states, probably in part because of the very warm winter (table 19.1). January and February temperatures were up to 5°F above average in Connecticut and New Hampshire. Not only does this add warmer months to the yearly average, but it removes the two coldest months of the year. In some years, a warm summer was offset by a cold winter, producing an average year. That was not the case in 1998, when the warm summer followed a winter of record warmth. Other El Niño years in the twentieth century do not stand out as being exceptionally warm from an annual perspective. The La Niña event of 1973 shows up in the instrumental record as being noticeably warmer than surrounding years, but it is not among the warmest of the century as is the El Niño–influenced year of 1998.

The evidence for the impact of volcanism on New England's climate is shown by cooler conditions in the year or two after several of the major eruptions in the twentieth century. Temperatures may not return to pre-eruption averages until four or five years after an eruption, if they ever return to those levels. One of the most distinct cases where this holds true is the period following the 1991 eruption of Mount Pinatubo in the Philippines. That was the largest climatically effective eruption of this century, and the cool summer of 1992 helped make that year the coldest of the 1990s in every New England state. Interestingly, 1992 was by far the coolest summer in southern New England (Connecticut, Rhode Island, Massachusetts), but 1997 was as cool or almost as cool in northern New England (primarily Vermont and Maine). Notice that annual temperatures did not reach or exceed the high levels of 1990 and 1991 until 1998. This scenario is even more apparent in the record of summer temperatures. The coldest year of the century in several states—1903—was also influenced by volcanism, that is, the three major eruptions in 1902, the largest being the eruption of Santa Maria in Guatemala.

Another interesting aspect of the volcanic influence on New England's climate is that the period from the mid-1920s to the late 1950s and early 1960s lacked any major volcanic eruptions. The warming temperatures in that time period are undoubtedly influenced by many other factors, but the lack of explosive volcanic activity during these decades removed a cooling component from the system. This time period coincides with the mid-warmth of this century that we mentioned above. Incidentally, several climatically effective eruptions in the 1960s to 1970s could have enhanced the overall cool climate of that time period.

Although there is evidence of the cooling associated with most volcanic eruptions during this past century, contrary evidence may be a function of the variable controls on New England's climate. Katmai erupted in Alaska in June 1912. While Katmai ejected a larger volume of material than the Pinatubo eruption, it did not contain as much sulfur, so it was not as climatically effective in its cooling potential. Nevertheless, Katmai was large enough to cool the climate in the northern hemisphere, as many temperature records show. This was not the case for New England. In fact, 1913 was the warmest year on record in Maine and second-warmest in Connecticut. It was not a noticeably cool year in any New England state. The impact of Katmai still should have been felt to some degree in 1913, but that does not appear to have been true here. We are not sure why 1913 was such a warm year, but perhaps prevalent circulation patterns overrode any volcanic cooling. Also, Katmai was a mid-latitude eruption while the others that we have discussed were equatorial eruptions. It is possible that New England's climate responds differently to mid-latitude eruptions than to equatorial ones.

Instrumental Records of Past Precipitation Trends

Precipitation trends show a true north-south difference. The 2001 analyses by the New England Regional Assessment Group indicate that southern New England precipitation increased over the last century by over 25 percent in all three states, while northern New England precipitation decreased by less than 5 percent for Vermont and New Hampshire, and by 12 percent in Maine (fig. 19.4).

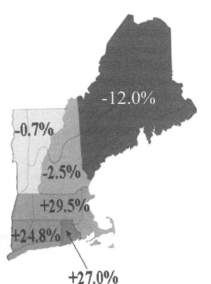

Fig. 19.4. Suggested change in mean annual precipitation, expressed as a percentage, for each New England state between 1895 and 2000. Values were determined in the same manner as those for MAT on figure 19.1 and explained in figure 19.2. Modified from the New England Regional Assessment Group. 2001. *Preparing for a Changing Climate: The Potential Consequences of Climate Variability and Change*. New England Regional Overview. U.S. Global Change Research Program, 96 pp., University of New Hampshire.

The three northern states, and particularly Vermont and Maine, show an over-all decrease from the beginning of the century to the serious drought of the mid-1960s. This trend influences the century-scale trend suggested by placing the best-fit line onto the data. The southern New England states show overall fluctuating conditions in the first half of the century, with noticeably low pre-cipitation totals in the 1910s extending somewhat into the 1920s in southern New England and Vermont. All of New England has been experiencing an increase in precipitation, on average, since the 1960s or 1970s. Ironically, 2001 is now the driest on record, especially in northern New England!

Several factors could explain a drying trend through the first half of the cen-tury in northern New England but not southern New England. The most obvi-ous mechanism would be a difference in storm tracks then compared to now. Storm tracks moving up the coast may have moved off shore before reaching northern New England more readily than they do now. This would prevent some of the more mositure-laden coastal storms from bringing rain or snow to the north. Another possibility would be that more tropical systems affected southern New England, although frequent tropical systems did not affect New England in the early part of the century (fig. 18.1). We believe changes in circulation patterns were the major reason for southern New England being wetter overall in the first half of the twentieth century.

In Figure 7.5, we show annual snowfall trends over the past seventy years for three stations across New Hampshire. These are an overall representation of trends across the region. In addition, we summarized the number of snow events with 1 inch or more and snow events with amounts of 10 inches or more as in table 7.2. In chapter 7, we defined a snow event as at least 1 inch of snow falling on successive days during a single storm, or from two or more storms that hit an area on successive days. From a human perspective, fast-moving storms that produce snow on successive days probably are not much different than a single storm over two days; either way, we have to shovel each day! Later in this chapter, we discuss changes in the number of snow events between the period of time covered by the instrumental record and that beyond the instrumental record.

Snowfall amounts are quite variable, as we have discussed before, and that remains true throughout the seventy years of record. While there is much year-to-year variability, the late 1950s to late 1970s stand out as the snowiest period of record. Other years have accumulated high snowfall totals, such as 1933 and 1934 and the record-setting winter of 1995/1996, but no time period really com-pares in overall snowiness. Yes, indeed, we had more snow to go through to school back in those days; it was not just that we were younger and smaller looking at big snow banks. The 1930s may be the next period of overall snowi-ness, but again, there is much year-to-year variability. The great variability in year-to-year snowfall is reflected by the standard deviation of annual snow shown in table 11.1. Annual snowfall in any two out of three years in Durham, New Hampshire, could be between 35 and 75 inches. In fact, coastal areas are more susceptible to inter-year variability, given the fickle nature of coastal storms, and particularly the track they may take. We explained these details in chapter 15.

Historical Records of Climate Trends

Earth's climate system is very dynamic and changes constantly, as we discussed in part II of this book. To understand how a region such as New England's climate behaves, it is important to study and evaluate climatic change beyond the intrumental record of the last one hundred years. One method for exploring how New England's climate has varied over the last few hundred years is to look at the historical record. Many, many individuals in the 1800s, 1700s, and even in part of the 1600s diligently recorded day-to-day weather conditions in their personal diaries. This was especially true of farmers, who depend on climatic conditions for their livelihood. In addition, clergy, the learned, and other individuals interested in their surroundings were very good at recording temperature conditions, often with descriptors such as hot, warm, cool, cold, rather than thermometer readings; precipitation type and often quantity, especially for snowfall; wind strength and direction; and, when barometers became available, pressure readings. These individuals also may have made notes on when particular plants bloomed or were harvested—that is, phenological information—and dates of ice-out on lakes and rivers, which often provide useful indicators of temperature variability from year to year. Notes on first and last frosts can be used to show how the growing season has changed with time. Snow depths and other conditions that affect travel were frequently recorded, providing a history of day-to-day weather, and thus of climate for the past few centuries. These records are enhanced by information from many other written records, such as newspapers, business journals, whaling logs, and other types of annals and summaries.

In his 1976 book on New England weather, Ludlum explored how the written record could be used to describe weather conditions during various battles and time periods during the Revolutionary War. Ludlum showed that information on wind directions, precipitation types, and changes in temperature could be used to show where high and low pressure systems and associated fronts were situated at particular times during a battle. Basically, one can create daily weather maps for the last few hundred years using this information. This information is beneficial not only to climatologists, but to historians as well. Besides, it is very enjoyable to read how individuals lived over the past few hundred years in New England, and how weather affected them every day just as it does today. We now give a few examples of how various aspects of our climate varied in the 1600s to 1800s.

Past Temperature and Precipitation. Robert Baron and David Smith compiled a large number of written records from New England and other parts of the Northeast during the 1980s. They published together and independently many scientific articles on their findings. One such article summarized changes in New England temperature and precipitation from 1640 to 1820 (fig. 19.5). In this example, Baron (1992) used the descriptive terms individuals used for seasonal and annual weather conditions to develop a numerical index that takes into account the amount of departure from what the diarist perceived as normal conditions. This scheme used a scale on which +14 would be "very hottest," –14 would be "very coldest," and 0 would be normal. Using this scheme, much of this 180-year time period would be classified as somewhat below average in temperature. In particular, the period from about 1750 to 1800 was especially cold for the colonists. This was a time when they were trying to establish themselves as a country, as well as trying to establish their farms, homes, and businesses. Weather conditions made this time period even more challenging. In some periods, such as around 1660, the 1740s, and during parts of the earliest 1800s, conditions were perceived as warmer than normal.

Baron's scheme for precipitation was similar, with +10 indicating the wettest conditions and –10 indicating the driest (fig. 19.5). Interestingly, much more year-to-year variability in precipitation was described, with the period from 1700 to 1720 being the only time frame over this 180-year period when precipitation stayed on one side of average, in this case, wetter than average. The mid-1600s were probably a little wetter than average, but alternating periods of wet and dry seem more prevalent than do alternating periods of hot and cold.

Past Growing Seasons. Perhaps one of the most useful applications for evaluating climatic change is in determining how changes in climate affect agriculture, and thus the ability for a community and its individuals to sustain themselves. Baron and Smith (1996) used the written record of spring and fall killing frosts to calculate growing season length for each of the climatic zones in each state. We give four examples in figure 19.6. Their records went from the 1600s, as in western Vermont, or from the 1800s, as in northern Maine, to 1948. Beginning in 1948, temperature records rather than field observations began to be used to indicate frosts. Consequently, direct comparisons to modern spring or fall freeze dates and growing season are not reliable (see figs. 8.3, 10.3, and 10.5). The four examples given in figure 19.6 show that there has been a great deal of variability in both freeze dates and the resulting growing season. The only very obvious trends are the decrease in growing season length in northern Maine during the 1820 to 1840 period (a result of both earlier fall freezes and later spring freezes) and the increase in growing season length in

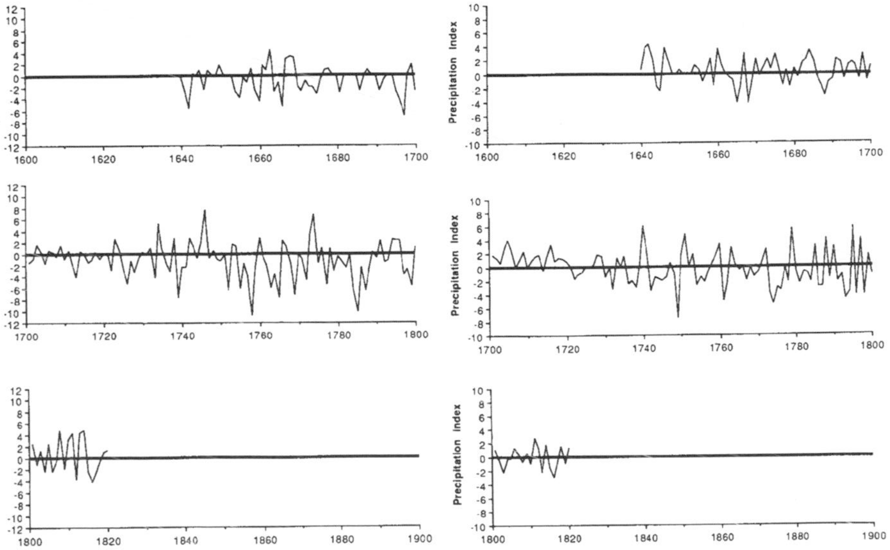

Fig. 19.5. Reconstructed New England temperature (*left*) and precipitation (*right*) for the period 1640 to 1820 using the written record and early thermometers from across the region. See text for description of the process used. Modified from Baron (Historical climate records from the northeastern United States, 1640 to 1900, Bradley and Jones, *Climate Since A.D. 1500*, Routledge, 1992; figs. 4.5 and 4.6, pp. 83–84). Used with permission from Taylor & Francis Books Ltd.

coastal Massachusetts beginning about 1890. Longer growing seasons in coastal Massachusetts primarily seem to be a function of later fall freezes. Interestingly, Baron and Smith's comparisons of growing season length from the late 1700s and early 1800s to the 1930s and 1940s show that all of New England experienced longer growing seasons except for the northern and southern interior climatic zones of Maine. The primary cause for the decrease in Maine was an overall later spring freeze in the 1930s to 1940s record. The areas of New England that experienced the greatest increase in growing season length were the coastal areas of Rhode Island, Massachusetts, and Maine. Like the instrumental records, these results show how differently Maine's climate has varied from other parts of New England as well as interior to coastal sections within the state.

Past Ice-out Dates. The ability to "walk on water" is of special interest to New Englanders, particularly from a recreational point of view. Ice-in and ice-out dates are important for determining when one can begin to think about ice skating or ice fishing on ponds, driving boats or snowmobiles across these same ponds, and other activities. Although ice-in does not necessarily mean that it is safe to venture onto frozen water bodies, it does mark the time when one can

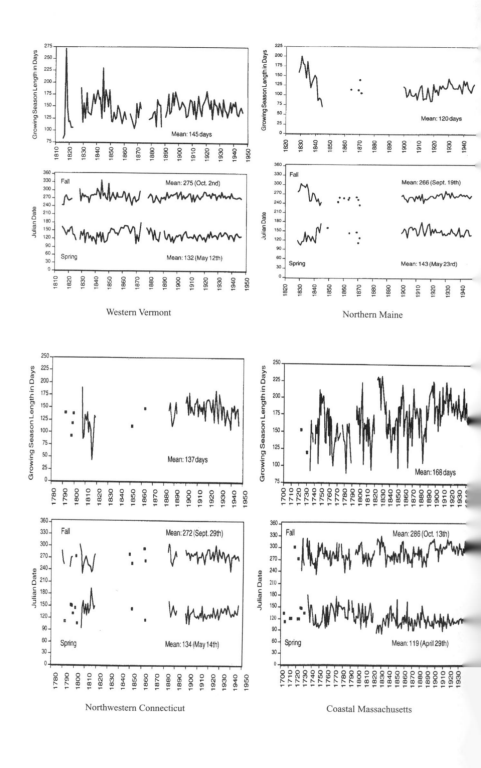

Western Vermont

Northern Maine

Northwestern Connecticut

Coastal Massachusetts

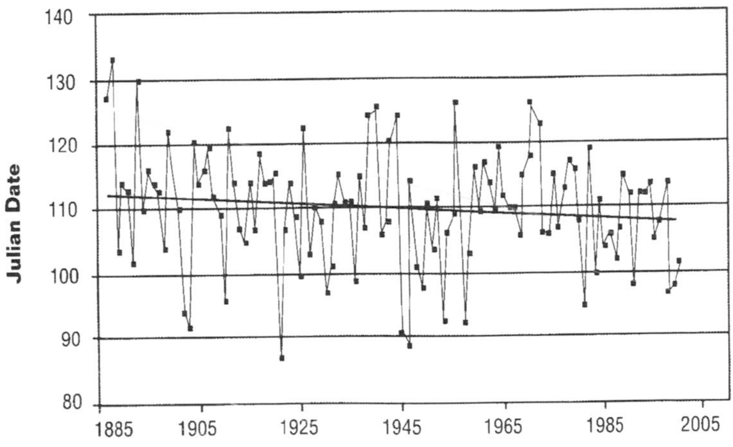

Fig. 19.7. Year-to-year variability in "ice-out" for Lake Winnipesaukee, New Hampshire. The best-fit line shown suggests that ice-out occurred about four days earlier in 1886 than at present. Ice-out for this lake occurs when the four ports on the lake are navigable by the steamship S.S. *Mount Washington*. From New England Regional Assessment Group. 2001. *Preparing for a Changing Climate*. New England Regional Overview. U.S. Global Change Research Program, 96 pp., University of New Hampshire.

begin to check ice thicknesses. At the same time, ice-out in the spring marks the time when boating season is getting closer. Records of ice-in and ice-out are useful indicators of spring and fall temperature variability, thus the length of winter conditions. Ice-out records have been kept for Lake Winnipesaukee since the 1880s, providing an indicator of past temperature changes for that part of central New Hampshire (fig. 19.7). The most distinct characteristic of this record for the past 115 years is the high degree of interannual variability. One should not be surprised by this after all of our previous discussions. Figure 19.7 indicates only one time period that showed consistency relative to the long-term average ice-out day. The period from the early 1960s to the late 1970s and early 1980s had ice-out dates that were close to or later than the average ice-out day of 20 April. The best-fit line to the data indicates that the average ice-out date is about four days later today than it was back in the late 1800s, but there is much variability in the data. Unfortunately, this makes for difficult planning for recreational use of Lake Winnipesaukee. On the other hand, unreliable ice-out dates epitomize New England's weather and climate.

Fig. 19.6. (*opposite*) Growing season length and timing of first and last frosts for northwestern Connecticut, coastal Massachusetts, western Vermont, and northern Maine using accounts recorded in the written record. Modified from Baron, W. R., and D. C. Smith (1996; figs. 6 and 7, p. 29, for Maine, figs. 19 and 20, p. 39, for Vermont, figs. 27 and 28, p. 46, for Connecticut, figs. 31 and 32, p. 49, for Massachusetts) "Growing Season Parameter Reconstructions for New England Using Killing Frost Records, 1697–1947." *Maine Agricultural and Forest Experiment Station Bulletin* 846.

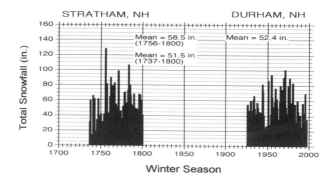

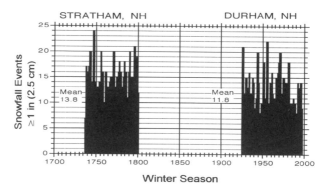

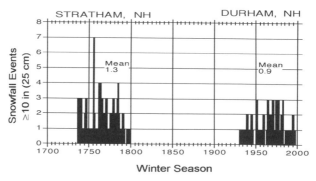

Fig. 19.8. Comparison of winter snowfall (*top*), number of snow events greater than or equal to 1 inch (*middle*), and number of snow events greater than or equal to 10 inches (*bottom*) for Stratham, New Hampshire, 1756–1801, and for Durham, New Hampshire, 1926–1997. Data for Stratham come from the diary of Samuel Lane, who recorded the weather daily, including days when it snowed and the amount that fell. His record extends from the winter of 1734 to that of 1800. He recorded snowfall totals for every storm after 1756 and 80 percent of the snows from 1737 to 1755, thus the actual average for the entire record could be greater than the 51.5 inches indicated on the top figure. A snow event means that snow was recorded on one day and any successive days after that to account for the possibility of snow beginning on one day and ending on the next day or for the possibility of several storms on successive days. From a human perspective, two different storms on successive days probably would be thought of as one storm.

Past Snowfall. We showed above that the founding fathers and mothers of New England probably had to endure colder temperatures, particularly in the latter part of the 1700s. Did they also have to brave more snowfall than we do at present? Fortunately, we can provide relatively close comparisons to the present. Samuel Lane of Stratham, New Hampshire, recorded daily weather events from 1734 to 1801. He was exceptionally diligent about recording snow events and taking measurements of the amount of snow that fell (fig. 19.8). In fact, from 1756 to 1800, he recorded a snowfall amount with each storm. He recorded 80 percent of the snowfall amounts from 1737 to 1956, a high percentage in its own right. We compared the snowfall records of Samuel Lane in Stratham to the recent snowfall records at Durham, New Hampshire (fig. 19.8 and table 7.2). It appears that the late 1700s were snowier than the present, possibly by as much as several inches per winter season, especially since single measurements taken during a snowstorm such as Samuel Lane recorded usually underestimate snowfall totals. Also, Durham records are not as complete as some of the other climatic stations in New England; the 57 inches of snow per year may be slightly high compared to the recent few decades (see fig. 7.5). Average annual snowfall for the period 1970 to 1999 is closer to 52 inches per year in Durham. This is only one record from the 1700s compared to only one modern station, but it does provide an interesting point to start discussions. The evidence suggests that the increase in snowfall in the 1700s may be due in part to a greater number of snow events per season (possibly two per season), as well as a greater number of bigger storms (fig. 19.8). Is this conclusion valid? More research on the number of past snowstorms may answer this question.

Summary

Over the last few hundred years, New England's climate has displayed the same variability, especially from year to year and across the region, that we see in the changes of our day-to-day weather. Using the instrumental record of the last 107 years (1895–2000), the 2001 New England Regional Assessment Group suggests that there is a general warming in temperature for every state except Maine. They base their conclusion on the application of a best-fit line on the temperature data, which it is a rough way to evaluate past temperature change. The also used this procedure to evaluate long-term precipitation changes; however, there has been much year-to-year change within that time frame. In general, temperatures have increased by over 1°F for the five states other than Maine, with Rhode Island showing a greater-than 2°F increase in mean annual temperature. Maine's temperature has decreased by 0.4°F over this same period, but it is not consistent across the state. The northern climatic zone of Maine has decreased by more than 1°F, the southern interior climatic zone has not changed, and the coastal zone has increased in temperature by slightly over 1°F. Changes in precipitation show north-south variability, with northern New England generally showing an overall decrease in precipitation and southern New England showing an increase in precipitation. However, more recently, annual precipitation for all of New England has been increasing, in general.

A longer record of New England climate with the same daily resolution provided by the instrumental record, may be obtained from historical writings such as personal diaries, journals, annals, logs, and newspapers. By assigning numbers to written descriptors of temperature (such as very hot, warm, cool, very cold) and of precipitation as compared to "normal" conditions (such as very wet or very dry), a lengthy record of New England temperature and precipitation was developed. Much cooler conditions seemed to exist in the mid- to late 1700s, whereas other time periods were generally variable in temperature. Annual precipitation shows the same degree of interannual variability, with only the earliest 1700s being generally wetter than normal. Growing season length and spring and fall freeze dates from the past few centuries reflect similar conditions to those documented by the instrumental record. That is, most of Maine, and particularly northern Maine, showed a general shortening of growing season, while the rest of New England had longer growing seasons. Coastal areas, even in Maine, showed the greatest increase in growing season length. The late 1700s also appear to have been snowier. Thus, the early colonist had to face harsh weather conditions, probably colder and snowier than today, during the time they were trying to build their homesteasds and establish an independent country. A tough task made even tougher by the climate of the time!

Chapter 20

What Does the Future Hold?

I tell people that anybody anywhere has just as much chance of making an eighteen-month guess—just as much chance of being right.
—DON KENT, METEOROLOGIST

Predicting the future climate of New England is a daunting task. On long time scales, such as the cycles observed over the last two million years, Earth should soon move toward another glacial period. This hypothesis is based on the fact that Earth's climate has been in an interglacial period for approximately 10,000 years, which is the typical length of an interglacial. If the global climate does shift in this direction, we can expect increasingly cool summers and possibly more severe winters, with substantive growth of continental glaciers in the higher latitudes of North America. These ice sheets may eventually advance into New England, as they have in the past, though we will not see that in our lifetimes.

Rather than being concerned over cooling conditions and glaciers, we should be more concerned about the prospect of warmer climates induced by increasing greenhouse gases, including carbon dioxide, methane, and water vapor. Earth's measured temperature over the past century over both land and water areas shows a trend toward warmer temperatures. Globally, the top six warmest years since 1880 have all occurred in the 1990s, with 1998 and 1997 as the two warmest years on record. Precipitation over land areas around the world also appears to be increasing, and the year 2000 was the third-wettest year on record. Some scientists have shown that global precipitation has increased by approximately 6 percent over the past century.

Similar to what is reported for global averages, New England's climate also shows strong indications that temperature and precipitation are both increasing, but this is more evident in the coastal zone. As we discussed in chapter 19, it is necessary to evaluate long records from the past to understand what is now happening to our climate and to anticipate what may happen in the future. For instance, the New England Regional Assessment Group (2001) shows that temperature has risen by about 0.74°F over the past century for the combination of

New England and New York. On a statewide basis, Rhode Island has experienced the largest increase in temperature, with statistical analysis indicting a 2.4°F increase from 1895 to 1999 (fig. 19.1). Warming has also occurred in the other New England states, with the exception of Maine, which has actually cooled by 0.4°F over the past century. In detailing these trends with greater spatial resolution, we found that coastal regions seem to be warming the most (modest warming is evident along the coast of Maine), with the most recent warming appearing to begin in the 1980s. We surmise that this is related to increasing urbanization in coastal areas, creating larger and stronger urban heat islands at the coastal cities. In addition, there is evidence that the sea surface temperatures have also warmed in the area, which would further modify the climate of the coastal cities. Interestingly, the summer of 2000, overall, was quite cool in New England, showing how year-to-year variability is still inherent within an overall increasing (or decreasing) trend (fig. 19.2). However, the procedure used to obtain these values is rather simple and it does not tell the whole story.

Changes in precipitation have also shown an interesting pattern across New England (fig. 19.4). Overall, the region has experienced an increase in rainfall of 3.7 percent from 1895 to 1999. Southern New England has seen a remarkable increase in annual rainfall totals, with Connecticut, Massachusetts, and Rhode Island experiencing a 25 to 30 percent increase over the past one hundred-plus years. Vermont and New Hampshire have seen a modest decrease in rainfall over the same period, and Maine precipitation has dropped by 12 percent. However, since the early 1980s, Maine's annual precipitation appears to be increasing. Although these precipitation patterns are largely not understood, they likely are related to shifting storm tracks—that is, nor'easters and Alberta Clippers—as we detail in chapter 19.

Even though temperature and precipitation are on the increase on the global scale and in coastal New England, this is not complete proof that we, as a society, have caused these changes. One of the most frequently raised questions about global warming stems from our interpretation of past instrumental temperature records. These data potentially contain contamination from many sources, including new instrumentation, station relocations, and new observation techniques, to name a few. These sources of errors may account for the 1 to 2°F increase in temperatures over the past century, although other means of evaluating lengthy trends in climate suggest that humans very well may be influencing present-day climate.

Most predictions of future climate over the next century come from general circulation models (GCMs). These are sophisticated numerical models that are supposed to simulate the atmosphere for the purpose of long-range forecasting. There are many different versions. Regardless of the version, GCMs often agree that global temperature and precipitation should increase as concentrations of atmospheric greenhouse gases increase, but regional impacts remain unclear. It is important to note that these models suffer many of the same shortcomings inherent within models that are used by meteorologists to forecast day-to-day weather changes. Although regional predictions for New England vary among the various models, a warming of near 7°F appears to be a reasonable consensus with a doubling of present carbon dioxide levels (fig. 20.1; New England Regional Assessment Group, 2001). Debates will be ongoing as to the cause of recent climatic change, but there is no doubt that climate has and will continue to change with time. Moreover, it is essential that we gather as much information as possible from both the present and past to completely understand our ever-changing climate system.

Another problem with GCMs is that most extreme events (such as intense precipitation events, tornadoes, hurricanes, and nor'easters) are too small in scale for GCM recognition. Therefore, the GCMs are of limited value in predicting extremes. Though the GCMS are of little assistance, global warming has implications for future events in the region, although with mixed possibilities. First, global warming would likely translate into warmer global sea surface temperatures (SSTs). It was found that warmer SSTs are strongly correlated with increases in tropical storm frequencies, at least in the North Atlantic basin (Wendland, 1977), which may affect storm frequencies in the eastern United States. Similarly, Emanuel (1987) reports that hurricane intensity is likely to increase under warmer conditions globally. Given these results, in conjunction with the fact that we are developing land in vulnerable locations such as the coastal zone and floodplains, our vulnerability to such extremes is likely to increase. However, time series of annual hurricane frequencies over the past one hundred-plus years do not show any trend toward increasing frequencies, nor is there any such temporal pattern for landfalling hurricanes in New England. Likewise, hurricane intensities do not appear to be increasing throughout this century, as evidenced by the timing of the ten most powerful storms to strike the eastern United States since 1900. These events occurred in the following years in decending order of intensity: 1935, 1969, 1992, 1919, 1928, 1960, 1900, 1909, 1915, and 1961.

A similar trend—or lack thereof—exists for snow-producing nor'easters that have struck New England. No consistent increase appears in the number or the intensity of these snow-producing coastal storms, although Davis and Dolan (1993) suggest that there has been an increase in the frequency of higher magnitude nor'easters in recent time. Their work is based on the coastal impact of such storms, and not all of those are snow producers. Consequently, different results may be found by evaluating a different impact of the extreme event. Understanding how extreme events may vary in the future comes down to the same need as for temperature trends, that is, the more data we can evaluate over longer time periods, the better our predictive capabilities.

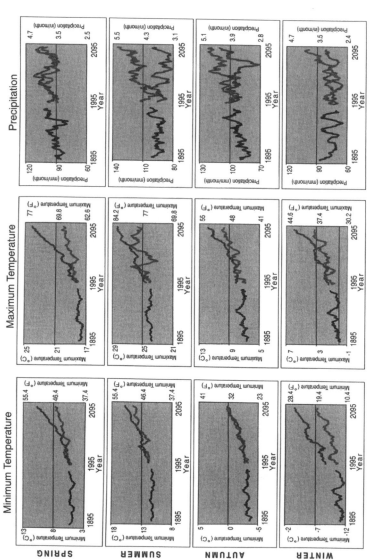

Fig. 20.1. Predicted changes in seasonal temperature and precipitation for New England using the Canadian and Hadley models (1994 to 2100) compared to the ten-year running means of historical gridded data. The Canadian model often predicts higher temperatures and lower precipitation amounts than the Hadley model. The exact numbers produced for each model are not as significant as the large differences between them. Modified from New England Regional Assessment Group. 2001. *Preparing for a Changing Climate: The Potential Consequences of Climate Variability and Change.* New England Regional Overview. U.S. Global Change Research Program, 96 pp., University of New Hampshire.

A second implication of global warming is based on the spatial dimensions of the warming. Most GCMs are predicting that higher latitudes will warm to a much larger extent than lower latitudes, and most of this warming should occur in winter. As a result, there would be a reduction in the temperature gradient between the tropics and the poles. It is this gradient, however, that drives most of the severe weather in the mid-latitudes, including in New England. As such, a gradient reduction may lead to a reduction in atmospheric mixing, thereby reducing specific types of severe weather, such as nor'easters. As mentioned above, the total number of East Coast nor'easters over the past fifty years appears to be decreasing, but the most powerful ones seem to be increasing in frequency. Not all of these storms had an impact on New England. To further complicate the equation, New England is positioned roughly halfway between the equator and the North Pole, and is exposed to both cold and dry air streams from the north and warm and moist air streams from the south. The interaction between these opposing air masses often leads to turbulent weather across the region. Also, because of the propensity of storm tracks to move across this region, the jet stream is frequently positioned overhead. The complicating factor here is that very small shifts in storm tracks and jet stream location lead to highly differing weather conditions region-wide. Currently, GCMs do not have the capability to predict how these storm tracks and jet stream locations may shift in a warmer climate.

Summary

Little is really known about the future climate of New England. Annual temperature and precipitation appear to have risen in the coastal region, though there is limited evidence that extremes have been on the rise. General circulation models forecast warmer conditions for the region, but are not yet sensitive enough to yield reliable information at a scale associated with most extreme events. At this point, we clearly need more time and research to assess the true impacts of a changing climate and how they may affect extreme weather phenomena. However, if global warming progresses further, there are likely to be winners and losers. On the up side, warming could lead to less-costly energy bills. On the down side, sea level is likely to rise, which would flood most beaches and coastal cities. Boston, Portland, Providence, and Cape Cod are just a few of the areas that could be affected in New England. In addition, New England might lose some of its colorful fall foliage because some tree species may disappear from the region. Also, the skiing industry is likely to be adversely affected by warmer winter temperatures. On the positive side, New Englanders will surely adjust to future changes as they have in the past.

A Retrospective

The only bad weather in New England is when we don't have any.
—DONALD HALL

Picture yourself waking up on a mid-April morning expecting to see beautiful sunshine, temperatures in the upper 60s or low 70s, a gorgeous spring day lying ahead. Although you may have to deal with "the mud," the warmth of spring awaits you. Unfortunately, as you wake, you are hit with the cold reality of New England's weather and climate. The snow is blowing on northeast winds, 6 inches have accumulated on the ground, several 2-foot drifts already bury the path between the house and the garage, and another 6 to 10 inches are predicted by nightfall. Alas, mud season will be extended another few weeks as this late-season snow melts.

This scene reflects what New England's weather and climate is all about. Winter keeps its hold on the region for as much as five or six months of the year; however, in some years, those beautiful spring or early summer days can come quite early. Similar warm days also can last well into November. For some individuals, a short winter may be a blessing, but others, such as those in the ski industry, say "bring on the snow into May." Nor'easters, such an integral part of New England culture, can be a blessing for some (a day off from school), but a curse to others (fishing boats remaining in harbor under gale-force winds). The fantastic aspect of New England's climate is that from season to season, and from year to year, no one is sure what the weather is going to be. Yes, unreliability is the epitome of New England's weather and climate.

Not only is the weather and climate so changeable among seasons and over long periods of time, but the several-hour drive from northern Vermont or New Hampshire to the Connecticut or Rhode Island coast can take one through many different weather conditions. It can be cold and snowy north of the notches, but beautiful, sunny, and in the 70s on the south shore of Connecticut, Rhode Island, and the Cape. New England is blessed with beautiful mountains and proximity to the ocean, but these features add much to the diverse weather conditions and climatic zones across the region. One need look no farther ahead than the approach of the first coastal snowstorm next winter to observe the complex nature of New England's weather and climate and the hard task we have predicting future conditions. Add the occasional tornado and land-falling hurricane (remember, on average, a landfall hurricane hits once every five years), and New England truly becomes a mixing bowl of weather and cli-mate not seen over such a small area anywhere else in the United States.

Through all of this, New Englanders take pride in how they cope with such conditions. Yes, New Englanders do talk about the weather, perhaps more then anyone elsewhere in the country, for it plays an integral part of their lives. Think about the industries that are so affected by the weather that exist in New England: farming, fishing and pleasure boating, forestry, maple sugaring, ship-ping, and orchard husbandry, to name just a few. A single storm or a lengthy period of hot or cold temperatures can change one's lifestyle. Also remember that the poor soil over much of the area is a direct result of a climatically driven factor, that is, the presence of glaciers over a mile thick over much of New Eng-land just fifteen to twenty thousand years ago, a very short time ago from a geological perspective. With the very short growing seasons, farming in the poor soils of the region is always on the edge, a fact that the earliest settlers probably became aware of all too quickly. Interestingly, climatic conditions in the 1600s and 1700s were overall colder and snowier than recent decades. In fact, Mark Lapping may have said it best in his lecture at Strawberry Banke, Portsmouth, New Hampshire, 1998, when he summarized the harshness of the New England climate and its impact on farming, fishing, and other activities necessary to survive. He suggested that if the United States had been settled from the west to the east, New England would more than likely be a wilder-ness area.

Despite the harshness of the New England climate, it supports a beautiful landscape. The wonderful sound of the peepers in the spring is so enjoyable during the evenings, although we must also mention the hatching of black flies later in that same season. Afternoon breezes off the lakes or ocean bless sum-mer's long days. The colors of the foliage season as autumn's temperatures tumble are unmatched anywhere in the world. Soft falling snow among the pine trees paints a very tranquil scene, while the awesome power exhibited by the blowing and drifting snow and crashing surf during a large nor'easter are awe inspiring. Indeed, New England's weather and climate have changed dra-matically with time and space across the region, and will do so in the future. New Englanders have adapted admirably in the past and they will continue to do so in the future. We are also quite sure that New Englanders will continue to talk about the weather and climate, passing down the memories of the extreme weather events they have endured. Weather and climate are surely a large part of the New England culture, past, present, and future.

Glossary

Advection: The horizontal transfer of any atmospheric property by the wind.

Aerosols: Tiny suspended solid particles (dust, smoke, etc.) or liquid droplets that enter the atmosphere from either natural or human (anthropogenic) sources, such as the burning of fossil fuels.

Air mass: A large body of air, usually 1,000 miles or more across, that is characterized by homogeneous physical properties at any given altitude.

Anticyclone: An area of high pressure around which the wind blows clockwise in the Northern Hemisphere and counterclockwise in the Southern Hemisphere.

Arctic front: The frontal boundary between cold arctic air and warmer air masses, usually lying to the south of it. It often coincides with the southern boundary of snow cover during the winter. Many depressions originate on it. In northwestern Canada during the winter months, for example, the frontal zone incorporates cold, dry, continental polar air and modified maritime arctic air from the Gulf of Alaska to the north of continental tropical air.

Backdoor cold front: A cold front moving south or southwest along the Atlantic seaboard of the United States.

Climatic singularity: A climatological or meteorological event that takes place on nearly the same day annually.

Condensation: The change of state from a gas to a liquid.

Convection: The transfer of heat by the movement of a mass or substance. It can only take place in liquids or gases. Motions are predominantly vertical and driven by buoyancy forces.

Coriolis effect: The deflective effect of Earth's rotation on all free-moving objects, including the atmosphere and oceans. Deflection is to the right in the Northern Hemisphere and to the left in the Southern Hemisphere.

Cumulonimbus: An exceptionally dense and vertically developed cloud, often with a top in the shape of an anvil. The cloud is frequently accompanied by heavy showers, lightning, thunder, and sometimes hail.

Cumulus: A cloud in the form of individual, detached domes or towers that are usually dense and well defined. It has a flat base with a bulging upper part that often resembles cauliflower. Cumulus clouds often indicate fair weather.

Cyclogenesis: The development or strengthening of middle latitude (extratropical) cyclones.

Cyclone: An area of low pressure around which the winds blow counterclockwise in the Northern Hemisphere and clockwise in the Southern Hemisphere.

Dew point (dew point temperature): The temperature to which air must be cooled (at constant pressure and constant water vapor content) for saturation to occur.

Easterly trade winds: Winds in the equatorial zone formed by air flow from the subtropical highs to the equatorial trough. They are northeasterly in the northern hemisphere.

El Niño: A significant increase in sea surface temperature over the eastern and central equatorial Pacific Ocean that occurs at irregular intervals, generally ranging between two and seven years.

ENSO (El Niño/Southern Oscillation): A condition in the tropical Pacific whereby the reversal of surface air pressure at opposite ends of the Pacific Ocean induces westerly winds, a strengthening of the equatorial countercurrent, and extensive ocean warming.

Equinox: The point in time when the vertical rays of the sun are striking the equator. 21 March is the vernal equinox or spring equinox in the Northern Hemisphere and 22 or 23 September is the autumnal equinox in the Northern Hemisphere. Length of daylight and darkness is equal at all latitudes at equinox.

Evapotranspiration: The combined processes of evaporation by which a liquid is transformed into a gas and transpiration by which plants transfer water to the atmosphere as water vapor.

Extratropical: Weather phenomena outside the tropics.

Freeze-thaw cycle: One oscillation during which the temperature drops below freezing, then rises back above freezing.

Freezing rain: Very cold rain that falls in liquid form, but freezes upon impact when striking an object or the ground.

Front: The boundary or boundary region that separates air masses of different origins and characteristics. Temperature gradients in any horizontal surface are large through the front. Different types of fronts are distinguished according to the nature of the air masses separated by the front, the direction of the front's advance, and the stage of development.

Greenhouse gas: A gas that absorbs long-wave radiation and therefore contributes to the greenhouse-effect warming when present in the atmosphere. The principal greenhouse gases are water vapor, carbon dioxide, methane, nitrous oxide, halocarbons, and ozone.

Hadley Cell: A thermal circulation proposed by George Hadley to explain the movement of the trade winds. It consists of rising air near the equator and sinking air near 30° latitude.

Insolation: The incoming solar radiation that reaches Earth and the atmosphere.

Jet stream: Relatively strong winds concentrated within a narrow band in the atmosphere.

La Niña: A condition where the central and eastern tropical Pacific Ocean turns cooler than normal; a component of the ENSO system.

Lapse rate: The rate at which an atmospheric variable (usually temperature) decreases with height.

Long-wave radiation: A term most often used to describe the infrared energy emitted by Earth and the atmosphere.

Meridional flow: A type of atmospheric circulation pattern in which the north-south component of the wind is pronounced.

Meteorological bomb: Rapid intensifying (deepening) of a mid-latitude cyclone at the average rate of 1 millibar per hour or 24 millibars per 24 hours.

Milankovitch Theory: A theory proposed by Milutin Milankovitch in the 1930s suggesting that changes in Earth's orbit were responsible for variations in solar energy reaching Earth's surface and climatic changes.

Ocean-effect snow: Snow produced when warm ocean water evaporates into colder air, quickly condenses into low clouds, and produces snow that is then pushed onto land by easterly winds.

Ozone: Ozone is composed of three molecules of oxygen (O_3) and as a result is a very strong oxidizer. It is photo-reactive, meaning it forms when nitrogen oxides and volatile organic compounds are exposed to sunlight. In high doses for long periods, it is harmful to plants, and causes respiratory problems in humans. It is the same chemical that blocks dangerous ultraviolet radiation from the sun high in the atmosphere, but at the surface, it is a principal component of urban smog. Ozone concentrations are measured in parts per billion (ppb), referring to the proportion of ozone in the air to the total volume of air.

Permafrost: A layer of soil beneath Earth's surface that remains frozen throughout the year.

Phenological information: Records of the seasonal occurrence of flora and fauna (dates of flowering, migration, etc.) and of the periodically changing form of an organism, especially as this affects its relationship with its environment, and as these are affected by climate.

Polar front: A semi-permanent, semi-continuous front that separates tropical air masses from polar air masses.

Polar jet: The jet stream that is associated with the polar front in the middle and high latitudes. It is usually located at altitudes between about 5 and 8 miles.

Pressure gradient: The rate of decrease of pressure per unit of horizontal distance. On a weather map, when the isobars are close together, the pressure gradient is steep. When the isobars are far apart, the pressure gradient is weak.

Radiational cooling: The process by which Earth's surface and adjacent air cool by emitting infrared radiation.

Relative humidity: The ratio of the amount of water vapor in the air compared to the amount required for saturation (at a particular temperature and pressure). The ratio of the air's actual vapor pressure to its saturation vapor pressure.

Rime: A white or milky granular deposit of ice formed by the rapid freezing of super-cooled water drops as they come in contact with an object in below-freezing air.

Short-wave radiation: A term most often used to describe the radiant energy emitted from the sun, in the visible and near-ultraviolet wavelengths.

Sleet: A type of precipitation consisting of transparent pellets of ice .2 inches or less in diameter.

Solar radiation: Short-wave radiant energy emitted from the sun.

Solstice: The point in time when the vertical rays of the sun are striking either the Tropic of Cancer (summer solstice in the Northern Hemisphere) or the Tropic of

Capricorn (winter solstice in the Northern Hemisphere). The solstice represents the day with the longest or shortest length of daylight of the year.

Stratosphere: The layer of the atmosphere above the troposphere and below the mesosphere (between 6.2 and 31 miles), generally characterized by an increase in temperature with increasing height.

Subpolar lows: Low pressure located at about the latitudes of the arctic and antarctic circles. In the Northern Hemisphere, the low takes the form of individual oceanic cells; in the Southern Hemisphere, there is a deep and continuous trough of low pressure.

Subtropical highs: Not a continuous belt of high pressure, but rather several semi-permanent anticyclonic centers characterized by subsidence and divergence located roughly between latitudes 25 and 35°.

Subtropical jet: The jet stream typically found between 20° and 30° latitude at altitudes between about 7 and 9 miles.

Supercooled water: The condition of water droplets that remain in liquid state at temperatures well below 32°F.

Synoptic scale: The typical weather map scale that shows features such as high and low pressure areas and fronts over distances spanning about 50 to 500 miles.

Teleconnection: A linkage between weather changes occurring in widely separated regions of the world.

Temperature inversion: An increase in air temperature with height. Defines the boundaries of the stratosphere, but can also occur in the troposphere under specific conditions.

Trade winds: The winds that occupy most of the tropics and blow from the subtropical highs to the equatorial low.

Tropopause: The boundary between the troposphere and the stratosphere. Height of tropopause varies by latitude and season.

Troposphere: The layer of the atmosphere extending from Earth's surface up to the tropopause (about 6.2 miles above ground). Most weather takes place in the troposphere.

Westerlies: The dominant westerly winds that blow in the middle latitudes on the poleward side of the subtropical high-pressure areas.

Zonal flow: A wind that has a predominant west-to-east component.

References Cited

Alley, R. B. 2000. *The two-mile time machine: Ice cores, abrupt climate change, and our future.* Princeton: Princeton University Press.

Baron, W. R. 1992. Historical climate records from the northeastern United States, 1640 to 1900. In *Climate since A.D. 1500,* ed. R. S. Bradley and P. D. Jones, 74–91. New York: Routledge.

Baron, W. R., and D. C. Smith. 1996. Growing season parameter reconstructions for New England using killing frost records, 1697–1947. *Maine Agricultural and Forest Experiment Station Bulletin* 846.

Barry, R. G. 1981. *Mountain weather and climate.* New York: Methuen.

Barry, R. G., and R. J. Chorley. 1998. *Atmosphere, weather and climate.* 7th ed. New York: Routledge.

Blue Hill Observatory. 1998. The storm. *Blue Hill Observatory Bulletin* 16 (Summer/Fall).

Calef, W. 1950. The interdiurnal variability of temperature extremes in the United States. *Bulletin of the American Meteorological Society* 31: 300–302.

Cember, R. P., and D. S. Wilks. 1993. *Climatological atlas of snowfall and snow depth for the northeastern United States and southeastern Canada.* Northeast Regional Climate Center, Publication No. RR 93-1, Cornell University, Ithaca, N.Y.

Climate Change Research Center. 1998. *New England's changing climate, weather, and air quality.* Durham: University of New Hampshire.

Danielson, E. W., J. Levin, and E. Abrams. 1998. *Meteorology.* Boston: WCB/McGraw-Hill.

Davis, R. E., and R. Dolan. 1993. Nor'easters. *American Scientist* 81: 428–39.

Davis, R. E., R. Dolan, and G. Demme. 1993. Synoptic climatology of Atlantic Coast north-easters. *International Journal of Climatology* 13: 171–89.

DeGaetano, A. T. 2000. Climatic perspective and impacts of the 1998 northern New York and New England ice storm. *Bulletin of the American Meteorological Society* 81: 237–54.

Dolan, R., and R. E. Davis. 1992. Rating northeasters. *Mariners Weather Log* (Winter), 4–11.

Emanuel, K. A. 1987. The dependence of hurricane intensity on climate. *Nature* 326: 483–85.

Garoogian, D. 2000. *Weather America 2001.* 2d ed. Lakeville, Conn.: Grey House Publishing.

Glickman, T. S., ed. 2000. *Glossary of Meteorology*. 2d ed. Boston: American Meteorological Society.

Grazulis, T. P. 1991. *Significant tornadoes, 1880–1989*. Vol. 1, *Discussion and Analysis*. St. Johnsbury, Vt.: Environmental Films.

Hauer, R. J., W. Wang, and J. O. Dawson. 1993. Ice storm damage to urban trees. *Journal of Aboriculture* 19(4): 187–94.

Henderson-Sellers, A., H. Zhang, G. Berz, K. Emanual, W. Gray, C. Landsea, G. Holland, J. Lighthill, S-L. Shieh, P. Webster, and K. McGuffle. 1998. Tropical cyclones and global climate change: A post-IPCC assessment. *Bulletin of the American Meteorological Society* 79: 19–38.

Keim, B. D. 1998. Record precipitation totals from the coastal New England rainstorm of 20–21 October 1996. *Bulletin of the American Meteorological Society* 79: 1061–67.

Kocin, P. J. 1983. An analysis of the "Blizzard of '88." *Bulletin of the American Meteorological Society* 64: 1258–72.

Kocin, P. J., and L. W. Uccellini. 1990. *Snowstorms along the northeastern coast of the United States: 1955 to 1985*. Boston: American Meteorological Society.

Leathers, D. J. 1994. A tornado climatology for the northeastern United States. Northeast Regional Climate Center, Publication No. RR 94-2, Cornell University, Ithaca, N.Y.

Leathers, D. J., B. Yarnal, and M. A. Palecki. 1991. The Pacific/North American teleconnection pattern and United States climate. Part I, Regional temperature and precipitation associations. *Journal of Climate* 4: 517–28.

Leathers, D. J., A. J. Grundstein, and A. W. Ellis. 2000. Growing season moisture deficits across the northeastern United States. *Climate Research* 14: 43–55.

Ludlum, D. M. 1963. *Early American hurricanes 1492–1870*. Boston: American Meteorological Society.

Ludlum, D. M. 1966. *Early American winters 1604–1820*. Boston: American Meteorological Society.

Ludlum, D. M. 1968. *Early American winters II 1821–1870*. Boston: American Meteorological Society.

Ludlum, D. M. 1970. *Early American tornadoes 1586–1870*. Boston: American Meteorological Society.

Ludlum, D. M. 1976. *The country journal: New England weather book*. Boston: Houghton Mifflin.

Ludlum, D. M. 1969. *The Vermont weather book*. 2d ed. Montpelier: Vermont Historical Society.

Lutgens, F. K., and E. J. Tarbuck. 1998. *The atmosphere*. 7th ed. Upper Saddle River, N.J.: Prentice-Hall.

Mayewski, P. A., and F. White. 2002. *The ice chronicles*. Hanover, N.H.: University Press of New England.

McKenzie, A. 1984. *World record wind: Measuring gusts of 231 miles an hour*. Eaton Center, N.H.: Alexander McKenzie.

Moran, J. M., and M. D. Morgan. 1997. *Meteorology*. 5th ed. Upper Saddle River, N.J.: Prentice-Hall.

Morrill, R. A. 1977. Maine coastal flood of February 2, 1976. United States Geological Survey, Open-File Report 77-533.

New England Regional Assessment Group. 2001. Preparing for changing climate: The potential consequences of climate variability and change: New England overview. U.S. Global Change Research Program, University of New Hampshire.

Rebertus, A. J., S. R. Shifley, R. H. Richards, and L. M. Roovers. 1997. Ice storm damage to an old-growth oak-hickory forest in Missouri. *The American Midland Naturalist* 137(1): 48–61.

Rooney, J. F., Jr. 1967. The urban snow hazard in the United States: An appraisal of disruption. *Geographical Review* 57, 538–59.

Ross, T., N. Lott, and M. Sittel. 1995. White Christmas? National Climatic Data Center, Technical Report 95-03.

Schmidlin, T. W., B. E. Dethier, and K. E. Eggleston. 1987. Freeze-thaw days in the northeastern United States. *Journal of Climate and Applied Meteorology* 26: 142–55.

Schneider, S. H., ed. 1996. *Encyclopedia of climate and weather*. 2 vol. New York: Oxford University Press.

Simpson, R. H., and H. Riehl. 1981. *The hurricane and its impact*. Baton Rouge: Louisiana State University Press.

Sisinni, S. M., W. C. Zipperer, and A. G. Pleninger. 1995. Impacts from a major ice storm: Street-tree damage in Rochester, New York. *Journal of Aboriculture* 21(3): 156–67.

Stommel, H., and E. Stommel. 1983. *Volcano weather*. Newport, R.I.: Seven Seas Press.

Thomas, R. B. 2000. *The old farmer's almanac 2001*. Dublin, N.H.: Yankee Publishing Incorporated.

Trewartha, G. T. 1981. *The Earth's problem climates*. Madison: University of Wisconsin Press.

Twain, M. 1935. *The Family Mark Twain*. New York: Harper and Row.

Watson, B., ed. 1990. *New England's disastrous weather*. Camden, Maine: Yankee Books.

Wendland, W. 1977. Tropical storm frequencies related to sea surface temperatures. *Journal of Applied Meteorology* 16, 477–81.

Williams, J. 1992. *The weather book*. New York: Vintage Books.

Zielinski, G. A. 2002. A classification scheme for winter storms in the eastern and central United States with an emphasis on nor'easters. *Bulletin of the American Meteorological Society* 83: 37–51.

Other Suggested Readings

Aherns, C. D. 1994. *Meteorology today*. 5th ed. Minneapolis/St. Paul: West Publishing Company.

Balsama, J., and P. R. Chaston. 1997. *Weather basics*. Kearney, Mo.: Chaston Scientific, Inc.

Blue Hill Observatory. 1993. The blizzard of March 1993. *Blue Hill Observatory Bulletin* 11 (Winter/Spring).

Bluestein, H. B. 1999. *Tornado alley*. New York: Oxford University Press.

Bomar, G. W. 1995. *Texas weather*. 2d ed. Austin: University of Texas Press.

Elsner, J. B., and A. B. Kara. 1999. *Hurricanes of the North Atlantic*. New York: Oxford University Press.

Gelber, B. 1998. *Pocono weather*. Stroudsburg, Pa.: Uriel Publishing.

Junger, S. 1998. *The perfect storm*. New York: Harper Paperbacks.

Lamb, H. H. 1995. *Climate, history and the modern world*. 2d ed. New York: Routledge.

Ludlum, D. M. 1989. *The American weather book*. Boston: American Meteorological Society.

Ludlum, D. M. 1998. *National Audubon Society field guide to weather*. New York: Knopf.

Minsinger, W. E., and C. T. Orloff. 1992. *Hurricane Bob: August 16–August 20, 1991*. Milton, Mass.: Blue Hill Meteorological Observatory.

Perley, S. [1891] 2001. *Historic storms of New England*. Beverly, Mass.: Commonwealth Edition.

Thaler, J. S. 1996. *Catskill weather*. Fleischmanns, N.Y.: Purple Mountain Press.

Index

New England cities, towns, and other geographic locations in the region are found after each individual state.